T0186520

ASSESSMENT OF THE EFFECTS OF CLIMATE VARIABILITY AND
LAND USE CHANGE ON THE HYDROLOGY OF THE MEUSE RIVER
BASIN

VRIJE UNIVERSITEIT

UNESCO-IHE Institute for Water Education

Assessment of the effects of climate variability and land use change on the hydrology of the Meuse river basin

ACADEMISCH PROEFSCHRIFT

ter verkrijging van de graad Doctor aan
de Vrije Universiteit Amsterdam en
het UNESCO-IHE Institute for Water Education te Delft,
op gezag van de rector magnificus
prof.dr. L.M. Bouter
en de rector prof.dr. R. Meganck,
in het openbaar te verdedigen
ten overstaan van de promotiecommissie
van de faculteit der Aard- en Levenswetenschappen
en het UNESCO-IHE Institute for Water Education
op donderdag 7 september 2006 om 15.00 uur
in het Auditorium van UNESCO-IHE

door

Min Tu

geboren te Wuhan, China

promoter: prof. dr. S. Uhlenbrook
copromoter: dr. P. de Laat

Published by A.A. Balkema Publishers, a member of Taylor & Francis Group plc.
www.balkema.nl and www.tandf.co.uk

ISBN 0 415 41694 9 / 978-0-415-41694-8 (Taylor & Francis Group)

Abstract

Anthropogenic climate change may lead to intensification of the global hydrological cycle and to increased flooding risk of rivers across most of Europe. A series of extreme floods in the European rivers occurring in the last two decades have stimulated discussions about the possible effects of climate variability/change and human interventions in river basins.

The aim of this research was to improve our understanding of the hydrological response of a large river basin (the Meuse) to climate variability and land use change. The study area covers the French and Belgian parts of the Meuse basin (upstream of Borgharen, about 21,260 km^2). The following specific research aspects have been addressed: identification of the temporal changes in the discharge regime (particularly the flood regime); identification of the temporal changes in the precipitation regime; linkage between the observed precipitation pattern change and large-scale atmospheric circulation; assessment of the rainfall-runoff relations; identification of historical land use changes and hydraulic activities in the Meuse basin and assessment of their potential impacts (particularly from land use changes). To obtain supportive evidences and also to demonstrate effect of spatial scale, the investigation has been extended to a few selected tributaries. The research was based on a combination of statistical trend analyses and hydrological modelling.

Based on the reconstructed discharge record (1911–2000) for the "undivided" Meuse near Monsin, the annual average discharge appears to have been relatively stable during the 20th century. Trend analysis of the flood series (both annual/seasonal maximum series and peaks-over-threshold series) derived from the daily discharge record (1911–2002) for the Meuse at Borgharen indicates that the magnitude of winter (November–April) floods in the Meuse river has significantly increased since 1984, accompanied by the increased frequency of winter floods (e.g. those greater than 800 m^3/s) since 1979. Based on the reconstructed Monsin discharge record (1911–2000), the summer low flow of the Meuse river, defined as the summer (May–October) minimum 10-day moving average discharge, shows a distinct decrease since 1933.

Trend analysis of the basin-averaged precipitation record (estimated from seven Belgian gauging stations, 1911–2002) for the Meuse basin upstream of Borgharen reveals that there exist apparent changes in the precipitation amounts, precipitation events and extremes during the 20th century and most of the observed changes occurred in the last two decades. The annual precipitation amount (November–October) appears to have just significantly increased since 1980, accompanied by a tendency towards more frequent intense precipitation events (e.g. exceeding 10 mm/d) in the winter half-year (November–April).

The results of the synoptic-climatological analysis using the synoptic data (1911–2002) of the European atmospheric circulation patterns (*Grosswetterlagen* system) and the North Atlantic Oscillation (NAO) index illustrate that the precipitation pattern change in the Meuse basin since 1980 is a reflection of the fluctuation of large-scale atmospheric circulation (mainly rain-associated circulation patterns such as West cyclonic Wz, Southwest cyclonic SWz and Northwest cyclonic NWz etc.); the wetter winter (December–February) condition since 1980 is likely to be a consequence of the strengthened NAO that brings stronger westerly surface winds across the North Atlantic to Europe.

Analysis of the annual runoff/rainfall ratio suggests that the annual rainfall-runoff relation in the Meuse basin has experienced some change around 1932, causing a significant decrease in the runoff proportion. There is a difficulty in explaining such a change on the basis of precipitation variability in the area. The increased magnitude of winter floods in the Meuse river over the last two decades is largely affected by increased antecedent precipitation depth over consecutive days (e.g. ten days), and thus can broadly be ascribed to climate variability. The hydrological simulations of the daily discharge record for Monsin (1911–2000) using the semi-distributed conceptual HBV model demonstrated that variations in the observed discharge record of the Meuse from 1911 onwards can be largely explained by variations in the meteorological variables (mainly precipitation, temperature and potential evapotranspiration). Nevertheless, the decrease of the runoff proportion around 1932 can not be explained by the meteorological data used, the forest type change in the basin and the uncertainty of the discharge record for Borgharen/Monsin. Graphical analysis indicates that the marked decrease in the summer low flow of the Meuse river around 1933 can not be explained by variations in the meteorological conditions (mainly precipitation and potential evapotranspiration). A likely cause is related to the quality of the Borgharen/Monsin discharge record analysed.

During the 20th century, the most obvious land use changes in the Meuse basin (upstream of Borgharen) are forest type change (from deciduous forest to coniferous forest since the end of the 1920s), modernisation of forestry and agricultural management practices as well as rapid urbanisation (since the 1950s). In addition, many hydraulic engineering works and management measures have been taken place in the Meuse river network, e.g. operation of weirs for navigation, canalisation and water abstraction from the Meuse, construction of the reservoirs, modification of the river channel etc. The in-depth investigation based on both results of the statistical trend analysis and the HBV-simulations clearly demonstrated the overall effects of historical land use changes on the discharge regime of the Meuse river are marginal or statistically undetectable. The attenuating effects of hydraulic works and management measures along the Meuse river network are too small to significantly impact the long-time trend of flood discharges downstream. The assumption that land use changes upstream of Borgharen have substantially influenced the occurrence of recent floods downstream could not be justified.

Acknowledgements

The research work for this thesis was carried out at UNESCO-IHE Institute for Water Education in Delft, The Netherlands. I would like to express my deep gratitude to the Institute for providing me with all necessary facilities and nice working conditions. The fund supporting this research (under the Delft Cluster project no. 06.03.04) from RIZA in The Netherlands is also gratefully acknowledged.

I am deeply indebted to Prof. Dr. M.J. Hall, who acted as my former promotor during the period 2002–2004. His professional guidance, tireless dedication and high responsibility (even during his serious illness!) always kept me working in the right way. I felt very sorrow for his worsening health in 2004 and his unfortunate passing away in April 2005.

I am very grateful to Prof. Dr. S. Uhlenbrook, who took the responsibility of being my promotor from 2005 onwards. His supervision and encouragement came timely and pushed me in an appropriate way to complete the research work resulting in this thesis. I also learned a lot from his detailed reviews and critical comments during the preparation of the thesis.

I would like to express my deep gratitude to my co-promotor, Dr. P.J.M. de Laat. He offered me the opportunity of undertaking this research and supervised me throughout the whole research including translation of the Dutch summary in this thesis. Beside the work, his understanding and concern, which indeed reduced my lonely sense of life abroad, are unforgettable.

I want to express my warm thanks to Dr. M.J.M. de Wit at RIZA. His constructive suggestions and kind supports have been of great value for this research. He friendly gave a lot of help during my study and patiently reviewed the publications, discussion documents and the draft of this thesis.

I also wish to express my appreciation to all staff members who have helped and supported me during my study at UNESCO-IHE, particularly Dr. Y. Zhou. I sincerely thank the MSc students: Mrs. M.L.C. Agor, Ms. M.P. Calderón Hijuma, Mr. A.G. Ashagrie and Mr. A. Yusuf for their contributions to this research, and Mr. J.A. Guzman for developing the SPELL-stat program, which was very valuable for the analyses of the time series.

I owe special thanks to all those organisations who have provided the necessary data for this research, particularly RWS, KNMI and WRO in The Netherlands, KMI and MET-SETHY in Belgium, DWD, StUA and WVER in Germany, Météo-France and SCEES in France. I also wish to thank those people who have helped me in acquiring data and/or understanding data processing, particularly Mr. P. Dewil, Mr. P. Hulst, Mr. G. Pohland, Dr. L. Pfister, Dr. P. Reggiani, Dr. T.A. Buishand, Drs. R. Leander and Mrs. M. Stam.

I extend my special and sincere thanks to Dr. J-M. Stam at DWW, The Netherlands. She offered me a temporal job in a Dutch-Sino cooperation project to bridge the half year to my PhD defence.

Furthermore, I acknowledge the members of the reading committee: Prof. Dr. A.J. Hans Dolman (VU Amsterdam, The Netherlands), Prof. Dr. H.H.G. Savenije (TU Delft, The Netherlands), Prof. Dr. S. Demuth (BfG and University of Freiburg, Germany) and Dr. J. Kwadijk (WL|Delft Hydraulics, The Netherlands) for reviewing this thesis.

Last but not least, I owe my loving thanks to my husband and my daughter who stayed in China during my study abroad. Their understanding and spiritual support allowed me to continue. Finally, I want to express my great gratitude to my mother for her deep love and strong encouragement.

List of acronyms

BfG	Bundesanstalt für Gewässerkunde (Federal Hydrological Survey in Germany)
BUND	Bund für Umwelt und Naturschutz Deutschland (German Branch of Friends of the Earth)
CIFOR	Center for International Forestry Research
CORINE	Coordination of Information on the Environment
CRU	Climatic Research Unit at the University of East Anglia in UK
DGRNE	Direction Générale des Ressources Naturelles et de l'Environnement (Directorate-General for Natural Resources and Environment of the Walloon Region in Belgium)
DWD	Deutscher Wetterdienst (German Meteorological Survey)
EEA	European Environment Agency
FAO	Food and Agriculture Organization of the United Nations
ICPR	International Commission for the Protection of the Rhine
IPCC	Intergovernmental Panel on Climate Change
KMI	Koninklijk Meteorologisch Instituut van België (Royal Meteorological Institute of Belgium)
KNMI	Koninklijk Nederlands Meteorologisch Instituut (Royal Netherlands Meteorological Institute)
MET-SETHY	Ministère wallon de l'Equipement et des Transports-Service d'Etudes Hydrologiques (Unit of Hydrological Studies of the Ministry of Equipment and Transport in the Walloon Region of Belgium)
NAO	North Atlantic Oscillation
RIZA	Rijksinstituut voor Integraal Zoetwaterbeheer en Afvalwaterbehandeling (Institute for Inland Water Management and Waste Water Treatment in The Netherlands)
RWS	Rijkswaterstaat (Directorate-General of Public Works and Water Management in The Netherlands)
SCEES	Service Central des Enquêtes et Études Statistiques (Central Office of Statistical Surveys and Studies of the Ministry of Agriculture and Fisheries in France)
StUA	Staatliches Umweltamt Aachen (National Environment Office at Aachen in Germany)
UK	The United Kingdom of Great Britain and Northern Ireland
USA	United States of America
VW	Ministerie van Verkeer en Waterstaat (Ministry of Transport, Public Works and Water Management in The Netherlands)
WHM	Werkgroep Hoogwater Maas (Flood Defence Task Force for the Meuse)
WMO	World Meteorological Organisation
WRO	Waterschap Roer en Overmaas (Water Board Roer and Overmaas in The Netherlands)
WVER	Wasserverband Eifel-Rur (Water Board Eifel-Rur in Germany)

List of acronyms

Table of contents

1   Introduction

1.1 Problem definition

Global climate change induced by increased greenhouse gas concentrations has been widely accepted and is often cited as involving temperature rise and, consequently, precipitation pattern change. With the projected climate change, the flood risk of rivers generally will increase across most of Europe (Houghton *et al.*, 2001). In Europe, floods are the most common natural disasters and the most costly in terms of economic damage (EEA, 2001). During the last two decades, a series of extreme floods have occurred in the European rivers (e.g. the Glomma in 1995, Norway; the Rhine and the Meuse in 1993 and in 1995, Germany, France, Belgium and The Netherlands; the Oder, Morava and Danube rivers in 1997, Poland, Germany and The Czech Republic; the Rhine river in 1998, Germany and The Netherlands; the Elbe in 2002 and 2006, Germany and The Czech Republic; see http://www.em-dat.net). Concern about the risk of river flooding is increasing. Whether this phenomenon of frequent floodings in Europe (as well as in many other parts of the world) is a symptom of anthropogenic climate change is still debated within the scientific community (e.g. BráZdil *et al.*, 2005). Different opinions frequently refer to a reflection of natural climate variability and the consequence of human interference on the hydrological systems (e.g. Bronstert, 1995; Savenije, 1995; Corti *et al.*, 1999; Nachtnebel, 2003).

The Netherlands forms the delta of a number of rivers, of which the Meuse and Rhine rivers are the most important. Large parts of the areas surrounding the rivers are prone to frequent floods. Safety against flooding in The Netherlands has been given a high priority. Compared with the Rhine river for which snowmelt in the Alpine region plays an important role in flood generation, the Meuse river as an almost purely rain-fed river is very sensitive to climatic change and variability (Uijlenhoet *et al.*, 2001). The seasonal variability of the discharge regime of the Meuse river also appears to be more pronounced than the other rain-fed rivers (e.g. the Main, Weser, Havel, Neckar rivers) of comparable size located in the vicinity of the Meuse basin (De Wit *et al.*, 2001). During the last two decades, five (i.e. 1984, 1993, 1995, 2002 and 2003) out of the seven largest floods in the Meuse river recorded in the period 1911–2003 occurred. The damages caused by the 1993 and 1995 extreme floods are particularly dramatic (see Wind *et al.*, 1999). Figure 1.1 shows the 1995 flooding of the Meuse river. There was a widely held public perception that the frequency and magnitude of floods in the river have substantially changed during the 20th century. It is often assumed that rapid land use changes (e.g. urbanization, road construction and modern agriculture) in the upstream areas of the Meuse basin (partly) triggered or aggravated these recent floods events downstream. Flood protection along the embanked parts of the Meuse and Rhine rivers in The Netherlands is based on the design discharge obtained from the flood frequency analysis of the measured discharge data (Chbab, 1995). If the measured discharge record is subject to non-homogeneity, new extreme observations can substantially affect the calculated design discharges of the river. Chbab (1995) has demonstrated influence of both the 1993 and 1995 flood events on the design discharge of the Meuse river (at Borgharen) against floods with a statistical average annual exceedence frequency of once every 1250 years. In 2001, the design discharge for the Meuse river at Borgharen has been updated from 3650 m^3/s to 3800 m^3/s (Parmet *et al.*, 2001). Moreover, some section of the Meuse valley in the Dutch Province of Limburg has no dikes. Even moderate discharges in the river can cause inundations and damages (Middelkoop and Van Haselen, 1999). As to the cause of the floodings, it has also been argued (e.g. Middelkoop and Van Haselen, 1999) that changes in land use in the Meuse basin mostly affect floods of moderate magnitude and extremely high flows will only be influenced locally when peaks are generated by heavy summer rain. Nevertheless, the statements are not yet substantiated by scientific evidence. Therefore, for practical reasons, it is very important to learn whether the flood regime of the Meuse river has significantly changed and whether the change, if present, has been notably aggravated by human activities or is mainly a sign of climate variability/change.

Figure 1.1 The 1995 flooding of the Meuse river in the Dutch Province of Limburg (left, source: http://www.wldelft.nl) and the low flow situation of the Meuse river (right, source: http://www.geo.uio.no/drought/textbook).

For the 21st century, higher winter precipitation totals and, as a result, increased flooding probabilities in the Meuse and Rhine river basins are expected due to climate change (Pfister *et al.*, 2004). This poses a big challenge to flood risk management in the lowland and delta areas of the river basins. These areas have high populations, intensive land uses and increasing economic activities and hence are very vulnerable to flooding disasters. After recent floodings in Europe, it is generally accepted that the traditional flood defence strategy through increasing the level of the dikes against floodings is not sustainable (e.g. EEA, 2001). To cope with the potential effects of climate variability and change, improved flood risk management in river basins is needed. In response to the floodings of the Rhine and Meuse rivers in the mid-1990s, the large IRMA-SPONGE research programme (see http://www.irma-sponge.org) has been carried out, aiming at development of methods and tools to assess the impact of flood risk reduction measures and of land-use and climate change scenarios, in order to support the spatial planning process for the river basins (Hooijer et al., 2004). In The Netherlands, a new flood management policy has been developed, see "Space for the Rivers" (VW, 1996) and "Water Management Policy in the 21st Century" (VW, 2000), with emphasis on creating more space for the rivers and increasing the river's discharge capacity. In addition, at the river basin level, international cooperation between the riverine countries has led to the issue of two important action plans: Rhine Action Plan on Flood Defence (ICPR, 1998) and Meuse High Water Action Plan (WHM, 1998). In the Meuse basin area, the Dutch authority has inititated some pilot projects in the Border Meuse (Dutch: Grensmaas) area. The Maaswerken project is a good example of how the new policy on flood protection and water management is being put into practice (see http://www.maaswerken.nl/main.php). To reduce the risks of flooding, additional flood mitigation measures in the upstream Meuse basin are also important. This has been clearly reflected in two of three distinct "fictive" strategies for the Meuse basin in the INTERMEUSE project (see Geilen *et al.*, 2001) within the IRMA-SPONGE research program: the SPONGE-strategy (recovery and increment of the sponge effect in upstream areas by increased infiltration) and the RETENTION-strategy (increase of the retention capacities in upland areas through storing water in reservoirs etc., especially during peak discharges). However, the effectiveness of these strategies has not yet been elaborated in detail. To what extent changes in the headwater areas of the river basin affect the discharge regime of the downstream Meuse river is not well understood.

Flood management of the Meuse river basin can not be considered independently from other important river functions such as navigation, water supply, electricity production, ecosystem and recreation (e.g. Middelkoop *et al.*, 2004). These river functions are hampered during periods of low flows (see Figure 1.1). In 1995, Belgium (Flanders) and The Netherlands signed the "Meuse Discharge Treaty", an agreement about the amount of water flowing through canals and the Border Meuse. According to the Treaty, the discharge distribution of the Meuse should ensure that the Border Meuse has a discharge of at least 10 m^3/s (Jaskula-Joustra, 2003). With increasing economic activities and accordingly increasing water demands, the problem of water shortages with respect to socio-economic, ecological and environmental aspects in the downstream Meuse river during dry periods is likely to be more severe. This raises a particular concern on the future low flow situation of the Meuse river (e.g. Uijlenhoet *et al.*, 2001; Jaskula-Joustra, 2003). The projections for a few Belgian subcatchments in the Meuse and Scheldt river basins (e.g. Gellens and Schädler, 1997; Gellens and Roulin, 1998) suggest that subcatchments characterised by strong infiltration could be subject to positive evolution of the groundwater storage and of the baseflow, whereas subcatchments with predominant surface runoff could exhibit the reverse effect. For The Netherlands, it is not clear which impact the climate change would have on the low flow of the Meuse, considering the future probably different-direction trends of increasing evapotranspiration, less precipitation in the summer but more precipitation in winter and the characteristics of the Meuse basin area (Jaskula-Joustra, 2003).

In order to respond adequately to climate change in water management practices and optimize the river functions in relation to each other, a good understanding of the hydrology of the Meuse river basin is essential. The discharge regime is an important aspect, particularly in case of extremes. Extreme large discharges determine the boundaries of safety and extreme low discharges provide the thresholds for navigation and water supply. For nature conservation the range of the discharges and the fluctuation is of importance. Regarding the regional water balance, annual average discharges are fundamental. How the discharge regime of the Meuse river reacts to the changing climate/climate variability and land use change is a central question of interest to integrated water management at the river basin level.

1.2 General review

A general literature review is documented in this section. Sub-section 1.2.1 briefly introduces current knowledge on potential impacts of climate variability and change on precipitation and subsequently on river flows, particularly at the European continent. Sub-section 1.2.2 reviews research approaches to studying land use impact and hydrological effects of land use change.

1.2.1 Potential effects of climate variability and change

Both climate variability and climate change are frequently analysed in recent literature. It is very important to understand the difference between two concepts. According to the precise definitions given by Houghton *et al.* (2001), climate change refers to a statistically significant variation in either the mean state of the climate or in its variability, persisting for an extended period (typically decades or longer), while climate variability refers to variations in the mean state and other statistics (such as standard deviations, the occurrence of extremes etc.) of the climate on all temporal and spatial scales beyond that of individual weather events.

Instrumental and proxy observations show that the Earth's surface has warmed, notably in the last half of the 20th century. Most of the observed warming over the last 50 years is likely due to the increase in greenhouse gases arising from human activities (Folland *et al.*, 2001). With the projected global temperature rise, the global hydrological cycle will intensify. It is likely that annual precipitation will increase in the Northern Hemisphere mid- and high latitude regions and extreme weather events will become more common. Flood magnitude and frequency particularly in higher latitude regions in the Northern Hemisphere are anticipated to increase because of increased precipitation (Houghton *et al.*, 2001). The general trends in streamflows in Europe inferred from projected climate change are likely to be increased winter flows across much of Europe and reduced summer flows in southern Europe. The consequences of climate change for the variation of flow through the year are likely to be increased flood hazards across most of Europe and increased low-flow frequencies particularly in southern Europe (McCarthy *et al.*, 2001). Nevertheless, the current projections of future climate change (in particular the projections of changes in extremes) and its impacts on runoff and streamflow, particularly on regional or basin scale, are associated with considerable uncertainties (McCarthy *et al.*, 2001). The impacts of climate change on hydrology are usually estimated by defining scenarios for changes in climatic inputs to a hydrological model using the output of General Circulation Models. The spatial resolution of the climate models, roughly ~ 200 km, is often too coarse to simulate the impact of global change on (most) individual river basins, especially for the intensities and patterns of heavy precipitation which are heavily affected locally. Moreover, the greatest uncertainties in the effects of future climate on streamflow arise from uncertainties in climate change scenarios because of insufficient information about the future development of our earth's climate (McCarthy *et al.*, 2001).

Scientists have made tremendous efforts to search for empirical evidences of ongoing climate change. There are ample indications suggesting that during the 20th century the precipitation amounts have increased over a significant range at the Northern Hemisphere mid- and high latitudes (especially during autumn and winter), very likely accompanied by even more pronounced increases in heavy and extreme precipitation events. These increases also vary in space and time (Folland *et al.*, 2001). There have been marked increases in annual precipitation in the second half of the 20th century over northern Europe, with a general decrease southward to the Mediterranean (Folland *et al.*, 2001). In recent climate change research, more attention has been paid to variations of extremes. Frich *et al.* (2002) reported that a significant proportion of the global land area was increasingly affected by a significant change in climatic extremes during the second half of the 20th century. Indicators based on daily precipitation data show more mixed patterns of change but significant increases have been seen in the extreme amount derived from wet spells and number of heavy rainfall events. Alexander *et al.* (2006) offered the most up-to-date and comprehensive global picture of trends in extreme temperature and precipitation indices based on the results derived from global station data (1901–2003). Precipitation indices show a tendency toward wetter conditions throughout the 20th century. Klein Tank and Können (2003) investigated more than 100 meteorological stations in the European continent for the period 1946–1999 (a warming episode) and found that for the stations where the annual amount increases, the index that represents the fraction of the annual amount due to very wet days (defined by percentile threshold) gives a signal of disproportionate large changes in the precipitation extremes. In general, in mid- and high latitudes in the Northern Hemisphere, the observed increase in precipitation is qualitatively consistent with the trend inferred from most model simulations for future climate (Houghton *et al.*, 2001).

The fact that the global mean temperature has increased since the late 19th century and that other trends have been observed does not necessarily mean that an anthropogenic warming effect on the climate system has been identified (McCarthy *et al.*, 2001). A crucial question in the global-warming debate concerns the extent to which recent climate change is caused by anthropogenic forcing or is a manifestation of natural climate variability (Corti *et al.*, 1999). There is considerable uncertainty in the magnitude of natural climate variability. The mechanisms that link the process of global warming and other large-scale atmospheric forcings are not yet clear (BráZdil *et al.*, 2005). Across large parts of the world the changes in annual precipitation associated with global warming are small compared to those resulting from natural multi-decadal climate variability (McCarthy *et al.*, 2001). Corti *et al.* (1999) showed that recent climate change in the Northern Hemisphere can be interpreted in terms of changes in the frequency of occurrence of natural atmospheric circulation regimes. They conclude that recent Northern Hemisphere warming may be more directly related to the thermal structure of these circulation regimes than to any anthropogenic forcing pattern itself. Studies have shown that wetter-than-normal conditions over many parts of northern Europe and Scandinavia are linked to strong positive values of the North Atlantic Oscillation (Houghton *et al.*, 2001). In recent years, interest for studies on climatic change prior to the human-induced time period is growing rapidly, because observed climate variability in the historical data was primarily the result of natural forces, without important anthropogenic effects. An overview of the European researches in historical climatology is, for instance, given by BráZdil *et al.* (2005).

A number of studies have been conducted worldwide to investigate possible trends in recorded hydrological data (including floods and droughts) and most of them focus on the past several decades, e.g. Burn and Hag Elnur (2002) in Canada, Douglas *et al.* (2000) in USA, García and Mechoso (2005) in South America. In general, the trends found in the observed streamflows are consistent with those identified for precipitation (McCarthy *et al.*, 2001). However, even if a trend in streamflow of a river is identified, it may be difficult to attribute it to global warming because of other changes that are continuing in a catchment (McCarthy *et al.*, 2001). The detection of anthropogenically forced changes in flooding is difficult because of the substantial natural variability. The dependence of streamflow trends on the flow regime further complicates the issue (Milly *et al.*, 2002). Milly *et al.*

(2002) found that the frequency of great floods (with discharges exceeding 100-year levels) in large river basins (> 200,000 km^2) has increased substantially during the 20th century and the recent emergence of a statistically significant positive trend in the risk of great floods is consistent with model simulations for the anthropogenic climate change. Based on a global modelling study, Milly *et al.* (2005) concluded that a significant part of hydro-climatic change during the 20th century (average annual runoff, 1971–1998 vs. 1900–1970) was externally forced. Huntington (2006) made an extensive review of the observed trends in a number of hydrologic variables (e.g. precipitation, runoff, tropospheric water vapour, soil moisture, glacier mass balance, evaporation, evapotranspiration, and growing season length). In spite of an ongoing intensification of the water cycle, the empirical evidence to date does not consistently support an increase in the frequency or intensity of major storms and floods. In two recent studies conducted by Svensson *et al.* (2005) and Kundzewicz *et al.* (2005), the trend results of multiple station records worldwide (including the European data from Finland, Germany, UK, the Czech Republic etc.) do not support the hypothesis of ubiquitous increase of high flows. However, there seems evidence that maxima occurred considerably more frequently in the latter sub-period 1981–2000 than in the earlier sub-period 1961–1980. To understand the characteristics of natural decadal-scale variability in streamflow, an increasing number of studies have reconstructed considerably longer records from various proxy data sources (McCarthy *et al.*, 2001).

1.2.2 Effects of human activities (particularly land use changes)

Research approaches

Experimental approach   The traditional approach to studying land use impacts has involved the use of (either single, paired or nested) experimental catchments. Single experimental catchments are studied to determine the effects of catchment alterations on themselves. However, these experiments have value only in terms of comparisons with their own historical data sets. It has also been argued that the analysis of data from a single-catchment experiment is more informative because it relates streamflow to factors that influence it rather than to streamflow from another (control) catchment. Its principal disadvantage is the complexity involved and more analysis of data are required than the paired-catchment approach (Toebes and Ouryaev, 1970). The paired-catchment approach has been frequently applied in the areas with similar physio-geographical catchment conditions. One of the two experimental catchments is established as a "control" and its physical characteristics are kept as constant as possible throughout the pre-treatment calibration period and post-treatment evaluation period of the study for comparison with the other "treatment" catchment. In small experimental catchments, the paired-catchment approach is often considered the best method to compensate for climate variability (Brooks *et al.*, 1997). A disadvantage of this approach is that the "control" catchment may itself be subject to significant changes and then the statistical analysis results may not be valid (Toebes and Ouryaev, 1970). The nested-catchment approach is a variant of the single-catchment approach. Two or more subcatchments are deliberately modified to show the effects of changes within the subcatchments and within the entire basin. The additional value of this approach is that the differences between small and large basins can be studied. This is invaluable for translation of the results (Toebes and Ouryaev, 1970). Many experimental catchments are situated in forested areas and the hydrological effects of forests have been studied extensively. Due to some practical difficulties, most of the experimental catchments are not more than a few hectares in size and the controlled paired-catchment experiments are rarely sustained for decades. In large river basins, time-trend analyses are often conducted using existing data (e.g. Toebes and Ouryaev, 1970; Bosch and Hewlett, 1982; Sahin and Hall, 1996).

Statistical approach   In recent literature, detection of changes in time series of hydrological and climatological data has received a growing interest. Effects of climate change and land use change on streamflow can be so significant as to produce a non-stationary signal in observed discharge time series. Changes in an observation series often occur in the forms of jumps, trends (instability of the mean) and instabilities of variance. Sudden shifts are often related to changes in the gauging location and practices, while gradual changes may rise from effects of climate variability and land use changes (Dahmen and Hall, 1990). Kundzewicz and Robson (2004) provided a brief review of the methodology of change detection in time series of hydrological data, embracing stages such as preparing a suitable data set, exploratory analysis, application of adequate statistical tests and interpretation of results. Trend analysis requires long records, preferably long than about 50 years, because trends quantified from short records may only be part of the climate-linked fluctuations. Observed data should be quality-controlled. Meta-data, including station moves, changes in instrumentation, changes in the time of observations etc., are acquired for the interpretations. There are some practical difficulties in studying land use change impact using the statistical approach. The flow at a gauging station represents the integrated runoff upstream of the station. It is often difficult to separate and quantify the individual effects of different land use changes within the catchment area on the basis of observed evidences in the hydrological response (Brooks *et al.*, 1997). Furthermore, it is difficult to distinguish between effects of land use change, climate variability or simply altered measurement techniques (Kundzewicz and Robson, 2004).

Modelling approach   Studies of land use change impact on catchment hydrology have been frequently conducted by catchment modelling. The advantages and limitations of the modelling approach to simulation of land use effects have been discussed for instance by Refsgaard and Abbott (1996), Beven (2001b) and Bronstert *et al.* (2002). In practice, both conceptual (grey-box) and widely physically-based (white-box) hydrological models (as well as model types "in between") have been applied at different scales. Theoretically, the advantages of physically-based models in land use impact studies are clear (e.g. Refsgaard and Abbott, 1996; Beven, 2001a; Bronstert *et al.*, 2002). However, this type of model requires a large amount of detailed high-quality data which in many cases do not exist or are not easily accessible. Other important disadvantages are discussed extensively for instance by Beven (1996 and 2001a). Owing to these limitations, in practice physically-based models are often used for a limited physical system or for research purposes (Watts, 1996), with particular interest in prediction of effects of land use changes in catchment area and assessment of the spatial dynamics of the runoff generation processes. Recent examples of such applications include Fohrer *et al.* (2001, with the SWAT model), Niehoff *et al.* (2002, with the modified WaSiM-ETH model), De Roo *et al.* (2002, with the LISFLOOD model), Ott and Uhlenbrook (2004, with the TAC-D model). Conceptual models require relatively limited data, but models of this type must be parameterised, which is one of the most difficult aspects of applying conceptual models to a particular catchment. These models are usually considered to be over-parameterised, incorporating errors or strong simplifications in their structure and are sensitive to the optimisation criteria employed, or in many cases even physically unrealistic (e.g. Watts, 1996; Beven, 2001b). Many conceptual models are lumped or semi-distributed and thus do not allow for an accurate description of the effects of spatial variability within the catchment. The applications of conceptual models in assessment of land use impact can be found in for instance Schulze and George (1987, with the ACRU model), Gross *et al.* (1989, with the IHDM model), Buchtele (1993, with the SAC model), Lørup *et al.* (1998, with the NAM model) and Matheussen *et al.* (2000, with the VIC model). For large basins, hydraulic modelling of the channel processes may be involved together with hydrological modelling (e.g. Beven, 2001b). It is important to stress that differences between observed data and simulated output are subject to uncertainties of various sources, including: random or systematic errors in the input data such as precipitation, temperature and evapotranspiration etc. used to represent the input conditions in time and space over the catchment; random or systematic errors in the recorded data such as the river water levels, groundwater heads, discharge data or other data used for comparison with the simulated output; uncertainties and errors related to parameter values and model structure (e.g. Refsgaard and Storm, 1996).

Many modelling studies of land use impacts have concentrated on the vegetation change (i.e. afforestation or deforestation) itself. Due to the complexity of the processes involved, the effects of management activities in forestry and agriculture are difficult to simulate and their modelling results are associated with high uncertainty (Houghton-Carr, 1999; Beven, 2001b). Refsgaard and Abbott (1996) and Beven (2001b) have commented the status of practical application of hydrological models. Beven (2001b) stated that although there have been many studies that report predictions of the hydrological impacts of change, there are, as yet, no studies where catchment-scale predictions made before a change have later been verified.

Effects on mean flows

Afforestation and deforestation are two of the most important land use changes influencing the hydrological response of a catchment. Catchment experiments worldwide have demonstrated that substantially altering the type and extent of vegetative cover on a catchment can significantly affect the interception and evapotranspiration (ET) processes and, consequently, cause a change in the runoff volume. Generally, land use changes that reduce ET increase annual runoff from a catchment, whereas land use changes that increase ET decrease annual runoff. Coniferous forest, deciduous hardwood, brush and grass cover (in that order) have been found to have a decreasing influence on annual runoff of the source areas in which the land covers are manipulated (Calder, 1993; Brooks *et al.*, 1997). The degree of change in annual runoff from a catchment depends on the intensity and extent of land development. The generalised relationship based on catchment experiments worldwide is that a 10% reduction in coniferous forest (deciduous forest, shrub), being converted to grassland, causes an average increase of 40 mm (25 mm for deciduous forest, 10 mm for shrub) in annual runoff (Brooks *et al.*, 1997). Experimental results also indicate that effect of reductions in forest cover of less than 20% on streamflow may not be detectable (Bosch and Hewlett, 1982). Specific management practices in forestry and agriculture (e.g. road construction, soil compaction during logging, use of heavy vehicles, intensive grazing) can reduce the infiltration capacity of soils and in turn the flow of water through the soil profile, leading to an increase in the (surface) runoff volume on a catchment. Urbanisation tends to increase surface runoff volume due to increased impervious area or reduced infiltration capacity. In small urbanised catchments, discharge of the sewage water and storm water from municipalities through drainage systems to areas outside the catchment can significantly affect the water balance in the catchment (Newson, 1995; Brooks et al., 1997). The hydrological effects of reservoirs are typically re-distribution of the river flow within the year and increased evaporation from water surfaces.

Effects on flood flows

Land use activities may affect storm flow response and in turn flood peaks through changes in vegetation cover, soil infiltration capacity, conveyance system, (increased) erosion and sedimentation (e.g. Brooks et al., 1997). Bronstert et al. (2002) listed the potential impacts of land use changes on surface and near-surface hydrological processes (fluxes or storages) under "normal" conditions in humid temperature zones. Forests and forest soils have popularly been thought to influence the timing of streamflow by storing water during wet periods and releasing water during dry periods because of their high infiltration and soil moisture storage capacities, and hence reduce flood peaks. Conversely, deforestation is generally accepted to be a cause of increased flooding downstream. However, the above view is subject to debate. A strong criticism in the literature is that the existing "knowledge" of the relationship between the upland forest and downstream flooding is based largely on perceived wisdom or myth, rather than on science (e.g. FAO and CIFOR, 2005). The change in vegetative composition can affect the ET rates and in turn increase flood peaks in the early or late part of the wet season. However, flood peaks in the middle of the wet season are rarely significantly affected, because soil moisture at this time is usually high (e.g. Brooks et al., 1997). It is often the management activities associated with forestry which are more likely to influence flood response than the presence or absence of the forests themselves (e.g. Calder, 1998). Case studies on small-catchment scale show that the construction of forest roads can intensify peak runoff from forested areas significantly (e.g. Kiersch, 2000). For an extensively logged catchment of 149 km^2 in Washington, USA, La Marche and Lettenmaier (2000) reported that the simulated effect of forest roads on peak flows were roughly equivalent to those predicted for harvest cutting effects alone.

Many agricultural activities can be summed up by "intensification" of agricultural land, which often bring about negative impacts on flood conditions. In the literature, hydrological effects of drainage of agricultural land and mechanisation of farming receive particular concern. Field studies show that land drainage may increase or decrease flood peaks, depending on the soil properties and drainage characteristics (Ward and Robinson, 1990). The recent views are that the drainage of heavily clay soils (prone to surface saturation) generally results in a reduction in flood peaks for large and medium events, while on more permeable soils (less prone to surface saturation) the more usual effect of drainage is to intensify subsurface flows, leading to higher peak flows for a given volume of runoff (Houghton-Carr, 1999). Intensive use of farm vehicles can make the surface compact and thus reduce the infiltration capacity of soils, consequently leading to increased surface runoff and higher peak discharges (Brooks et al., 1997).

Urbanisation is often cited as one of the major human modifications to flood hydrology in developed areas. Houghton-Carr (1999) has summarized the generally recognised effects of urbanisation together with some other consequences that are not so widely recognised. For example, urbanisation typically leads to increases in flood peak discharge and total runoff volume, and a decrease in the travel time of flow to the catchment outlet. Urbanisation effects tend to be more pronounced in the response to small, short return period storms (which otherwise yielded low percentage runoff and little overland flow), than in the response to severe, high return period storms (which already have a typical urban high percentage runoff and increased overland flow). Also urbanisation effects tend to be less for widely impermeable catchments (which already have a typically-urban high percentage runoff and fast response) than for more naturally permeable catchment (which have a low runoff coefficient and slower response).

Generally, land use activities within a catchment area mainly influence the runoff generation, whereas river engineering and management measures along the river system influence discharge conditions (Bronstert *et al.*, 2002). Hydraulic activities directly influence natural flows of rivers and discharge conditions of channels (runoff routing), and thus often impact flood regimes. The effect of a reservoir (particularly for flood control purpose) is typically to lag and attenuate (i.e. reduce the amplitude, while maintaining the volume) the flood hydrograph from the catchment (Houghton-Carr, 1999). Artificial modifications to river channels are commonly performed in areas that are prone to flooding. The anticipated consequences are lowered water levels in the modified reach. However, channel modification may cause increases of flood peaks downstream (due to less retardation). The effects can be assessed by hydraulic-routing methods with the proper consideration of the interaction between the floods in the main channel and in the tributaries downstream (WMO, 1994).

Land use activities can have a noticeable effect on storm runoff volume, peak magnitude and timing of the peak for precipitation events that are not extreme in terms of amount and duration. Figure 1.2 illustrates a hypothetical annual hydrograph from a humid climate to demarcate the prevailing influences on the component parts of the river regime. It is generally assumed that changes in land use, particularly changes in forest cover, will more likely affect small- and medium-size floods for example with return periods of five year or less, whereas major floods in rivers are affected more by meteorological factors than by land use activities in upland catchments (Brooks *et al.*, 1997), which was, however, hardly proofed at river basin scale. Reed (1999) stated that from time-to-time and place-to-place, less dramatic land use changes are alleged to increase flood frequency; however, it is very difficult to verify or discount effects of land use changes on flood frequency.

Effects on low flows

Most studies of land use impacts have concentrated on floods and mean flows, since the impact on low flows and droughts seems to be regarded as less important. The effect of land use change on low flows during dry periods depends on competing processes, most notably changes in evapotranspiration and infiltration capacity (Calder, 1998). Each combination of both the dominating natural processes and the anthropogenic impacts has a different effect on the low flow regime. Therefore, the relatively quantitative impacts of various anthropogenic processes and factors on the low flow regimes vary substantially in different river catchments (Smakhtin, 2001). Forests' regulation of groundwater recharge is controversial. Upland forested catchments commonly are viewed as being important recharge zones for aquifers,

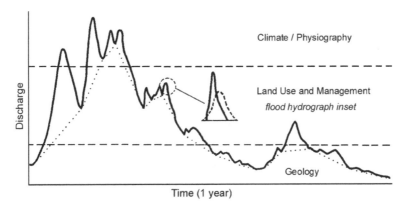

Figure 1.2 The annual hydrograph of a river (highly simplified) to illustrate the broad
controls on volume and timing of runoff. Dotted line indicates an extremely
simple separation of the baseflow, whose proportion of the annual flow total
constitutes a baseflow index (source: adapted from Newson, 1995).

because forests are often situated in areas with high annual precipitation and are
associated with soils that have high infiltration capacities. However, such occurrence
would be rare and would be significant only where small catchments feed a localised
groundwater aquifer (Brooks *et al.*, 1997). Field experiments as well as modelling
studies demonstrate that afforestation tends to reduce low flows relatively stronger
than annual flows. In urbanised catchments, low flows have a tendency to decrease
due to the effects (e.g. preventing the slow infiltration of water and in turn reducing
groundwater recharge) of urban impervious surfaces upon direct runoff, infiltration
and evapotranspiration. The lakes, reservoirs and wetlands in a river system can
attenuate the natural fluctuation in discharge. Low flows of rivers during dry periods
after regulation of the reservoir can either increase or decrease, depending on the
operational management of the reservoir (Smakhtin, 2001). Apart from indirect
anthropogenic impacts on low flow generating processes, there are human activities
impacting directly on low flows by removing water directly from or adding water to
the stream (Smakhtin, 2001).

Scale consideration

The influences of land use on hydrological processes are widely recognised (e.g.
Newson, 1995; Brooks *et al.*, 1997). Much of the knowledge has accumulated from
numerous field investigations in different environments. However, evaluation of the
hydrological effects of land use changes on catchment or river basin scale is difficult
(WMO, 1994; Brooks *et al.*, 1997). The net effect of various land use changes that
usually are spatially distributed and occur over time in catchments could be
cumulative or averaged-out. Hence transferring the results obtained on plot or small-
catchment scale to larger catchments is problematic (e.g. Brooks *et al.*, 1997;
Kiersch, 2000). It is more challenging to quantify the impact of land use change for
large river basins where the interactions between land use, climatic characteristics
and the underlying hydrological processes are often more complex and dynamic
(e.g. Uhlenbrook *et al.*, 2001). Scale consideration is of fundamental importance
when assessing the impact of land use changes on catchment scale. Blöschl and
Sivapalan (1995) have clarified the differences between the concepts of "process

scale", "observation scale" and "modelling (working) scale". The general views are that the effects of land use changes on hydrological regimes (e.g. mean flows, peak flows, base flow and groundwater recharge) will be most readily felt in smaller catchments of up to several hundred km^2, whereas in relatively large catchments (e.g. > 1,000 km^2), the cumulative catchment effects are often insignificant (e.g. Kiersch, 2000).

Flooding concerns are often restricted to the downstream areas of river basins. In large river basins, effects of land use practices in headwater subcatchments on peak flows tend to be offset due to time lag between different tributaries, different land uses and variations in storm rainfall (Kiersch, 2000). Moreover, major floods in large river basins often occur near the end of the rainy season when heavy rain falls in a number of subcatchments (simultaneously) and usually soils over large parts are already more or less saturated. Under such circumstances, the effects of different land uses make little sense (FAO and CIFOR, 2005). Due to the complexity of the runoff generation processes involved, the hydrological implications of extensive and long-term changes in vegetative cover are still controversial.

Generally, an increase in basin size increases the volume of the water-bearing rock and hence its storage capability. Moreover, an increase in basin size causes a levelling and smoothing of the various local factors. Therefore, with an increase in basin size the minimum flow tends to increase (Sokolovskii, 1971).

1.3 Research objectives and applied methodologies

1.3.1 Research objectives

The general aim of this research is to improve our understanding of the hydrological response of a large river basin (the Meuse) to climate variability and land use change. Five specific research objectives were defined:
- Identify the temporal changes in the discharge regimes (particularly the flood regime) of the Meuse river and a few selected tributaries;
- Identify the temporal changes in the precipitation regimes in the Meuse basin and the selected subcatchments;
- Explore the link between the precipitation variability in the Meuse basin and large-scale atmospheric circulation;
- Assess the rainfall-runoff relations and their temporal changes in the Meuse basin and the selected subcatchments;
- Identify historical land use changes and hydraulic activities in the Meuse basin and the selected subcatchments, and assess their hydrological impacts (particularly from land use changes).

This research emphasizes the relations between the observed streamflow changes and the observed precipitation changes in the Meuse basin over the 20th century. Based on this research, one can not judge whether the observed streamflow changes occurring in the past decades is a symptom of anthropogenic climate change. Therefore, the term "climate variability" is preferably used in the present thesis.

1.3.2 Applied methodologies

This research is based on a combination of the statistical trend analysis approach and the hydrological modelling approach. Statistical trend detection methods are applied to detect inconsistencies and non-homogeneities (both gradual trends and abrupt changes/shifts) in the data series of various variables (e.g. hydrological, hydro-meteorological and climatic variables). Because annual variability of hydrological time series for the Meuse is relatively high, the outcomes of linear trend tests may easily suggest absence of significant trends. Therefore, the identification of abrupt changes or shifts (regarding the mean values) in the data series is emphasised in the statistical trend analyses. Moreover, it is very useful when searching for possible reasons behind the changing hydrological behaviour of the river system. Comparatively, the detection of linear trends receives less attention. A description of the applied statistical test procedure is given in section 2.8.

Apart from the limits of currently available modelling techniques for evaluating land use impacts in large basins, the available data for this research is not sufficient to enable a detailed process-based modelling study of the land use change impact in the entire Meuse basin (upstream of Borgharen/Monsin). In this research, a hydrological modelling study using the semi-distributed conceptual model HBV is conducted on a daily time scale to gain insights into the effects of climate variation and the historical land use change in the Meuse basin.

Based on the statistical results in combination with the hydrological modelling results, the historical rainfall-runoff relations in the Meuse basin and the effects of human activities can be assessed. To obtain supportive evidences and also to demonstrate effect of spatial scale, the statistical investigation is extended to a few selected tributaries in the Meuse basin.

1.4 Structure of the thesis

This thesis is organised in eight chapters. It starts with an introduction in Chapter 1. In addition to the definition of hydrological problems, and research objectives and methodologies, a general review of current knowledge relevant to the research topic is also provided in this chapter. Chapter 2 describes the data used in this research and the statistical tests applied. An overview of historical land use changes and hydraulic activities in the Meuse basin as well as in the selected subcatchments is also provided. Chapter 3 presents the temporal changes identified in the discharge regimes of the Meuse river and the selected tributaries. Subsequently, Chapter 4 presents the temporal changes identified in the precipitation regimes in the Meuse basin and some of the selected subcatchments. To improve our understanding of precipitation variability in the Meuse basin, a synoptic-climatological analysis based on the *Grosswetterlagen* system and the North Atlantic Oscillation (NAO) index has been carried out and the results are provided in Chapter 5. Based on the statistical results and the catchment modelling study, Chapter 6 makes an assessment of the historical rainfall-runoff relationships in the study areas (emphasizing the entire Meuse basin) in the context of climate variability. The potential impacts of human activities on the discharge regimes of the Meuse river and the selected tributaries is addressed in Chapter 7. Chapter 8 gives an overall summary of this research and then the final conclusions are drawn. In addition, further researches are recommended in Chapter 8.

The outline of the thesis is given in Figure 1.3.

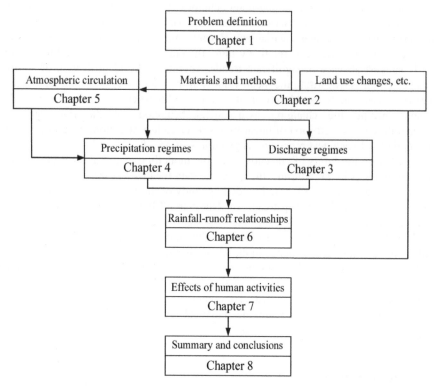

Figure 1.3 Structure of the thesis.

2   Materials and methods

2.1 Brief description of the Meuse basin

The Meuse (Dutch: Maas) river is one of the largest rivers in northwestern Europe. The river length measures almost 875 km from its source in France to Hollandsch Diep (the confluence with the Rhine river) in The Netherlands. The total basin area is about 33,000 km^2, including parts of France, Luxembourg, Belgium, Germany and The Netherlands (Figure 2.1).

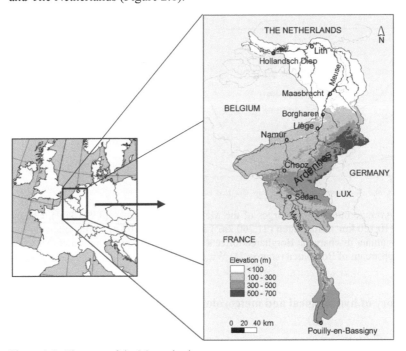

Figure 2.1 The map of the Meuse basin.

The elevations in the Meuse basin vary from less than 100 m in the lowland to more than 500 m in the Ardennes (see Figure 2.1). Three major geological zones can be identified: *i*) The Lorraine Meuse comprises the upper reach from the source at Pouilly-en-Bassigny to the confluence with the Chiers river near Sedan, mainly consisting of sedimentary Mesozoic rocks. *ii*) The Ardennes Meuse comprises the central reach from Sedan to the Belgian-Dutch border near Eijsden, transecting the Paleozoic rock of the Ardennes Massif. *iii*) The lower Meuse corresponds to the Dutch section of the river, formed by Cenozoic unconsolidated sedimentary rocks. The stretch downstream from Lith has tidal influence (Berger, 1992).

The Meuse basin has a temperate maritime climate, with a relatively cool summer and mild winter. The precipitation is evenly distributed throughout the year, predominantly frontal in winter and often convective in summer. The average annual precipitation is influenced by the differences in elevation, with the largest annual value (1000–1300 mm) in the Ardennes and less precipitation (700–800 mm) in the downstream lower area. The evaporation shows peak values during the summer months and low values during the winter months.

The Meuse river is generally classified as rain-fed river and has a rainfall-evaporation regime. Snow plays a very small but significant role in the runoff generation process. The river is characterised by a pronounced seasonal flow regime showing a year-to-year variability. Generally floods occur during winter and low flows occur during summer and autumn. The maximum flow can be as high as 3000 m^3/s and the minimum flow as low as 10 m^3/s. The long-term mean flow is about 250 m^3/s. The main tributaries include the Chiers, the Viroin, the Semois, the Lesse, the Sambre, the Ourthe, the Rur (not the Ruhr), the Niers and the Dieze (Berger, 1992). Figure 2.2 shows the average monthly discharges of different gauging stations along the Meuse river.

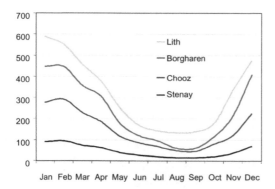

Figure 2.2 Average monthly discharges of the Meuse river at Seenay (3,904 km^2), Chooze (10,120 km^2), Borgharen (21,260 km^2) and Lith (29,370 km^2). The relatively low summer discharge at Borgharen reflects the consequence of canal water extraction upstream of Borgharen (source: De Wit *et al.*, 2001).

2.2 Inventory of hydrological and meteorological data

2.2.1 Water level/discharge data

Ideally the streamflow (used interchangeably with discharge) records used for evaluating the effects of land use changes and climate variability should cover the entire 20th century. Unfortunately, most streamflow gauging stations in the Meuse basin were initiated in the late 1960s or even later. In the FRIEND (Flow Regimes from International Experimental and Network Data) data base, from more than 30 stations in the basin, only several stations are suitable for trend analysis with regard to their record lengths of longer than 25 years. However, these records are not updated until present and some contain gaps of several years. Therefore, much effort was made to reconstruct long and uninterrupted discharge records.

As far as The Netherlands is concerned, the discharge of the Meuse at Borgharen (near Maastricht, see Figure 2.3) is normative. The drainage area upstream of Borgharen is about 21,260 km^2. Between Liège (Belgium) and Borgharen several canals (Figure 2.3) extract water from the Meuse (partly to areas outside the Meuse basin). In the case of low flows, the reconstructed record for the "undivided" Meuse

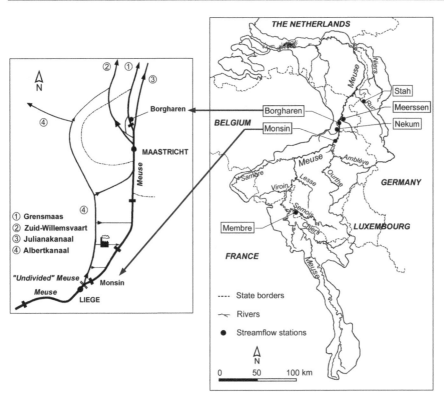

Figure 2.3 Overview of the discharge stations used in this study. The sketch-map of canals between Liège (Belgium) and Borgharen (The Netherlands) is modified from Berger (1992).

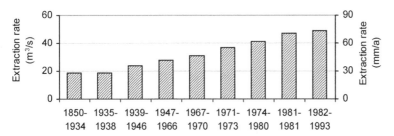

Figure 2.4 Estimates of water extractions from the Meuse between Liège and Borgharen. The bar diagram is made based on the data from Bos (1993).

near Monsin (Figure 2.3), a few kilometres north of Liège, is generally taken as the reference record. This reconstructed record (referred to as the Monsin record) was generated from the daily observations at Borgharen, corrected for water extractions (see Figure 2.4) from the Meuse between Liège and Borgharen (Bos, 1993). It is important to stress that the extraction estimates are not accurate especially for the period prior to 1990 and may roughly approach the reality only on an annual basis. For the Monsin record in this period (1911–1990), the correction factor is an annual mean value of abstractions throughout the year, with exception of the dry year 1976.

One would expect that more water may be extracted from the Meuse in dry periods/seasons if available, or less if not available. It should also be mentioned that almost the entire stretch of the Meuse river in the Ardennes is completely regulated through weirs. During periods of low flows, weirs are operated to maintain a minimum water level for shipping (Berger, 1992). The influence of the above mentioned extractions and regulation is relatively small during flood events, but can cause substantial day-to-day fluctuations during low flow periods that are not caused by rainfall-runoff processes. In this study, both the Borgharen record (1911–2002) and the Monsin record (1911–2000), provided by Rijkswaterstaat Direction Limburg, are analysed for the Meuse. The daily observations at Borgharen were recorded at 8:00 AM (local time) before 1975 and after that they were the average values of hourly observations per day.

A few gauging stations of the tributaries were also included in the trend analysis: Stah (Germany) on the Rur, Membre (Belgium) on the Semois, Nekum (The Netherlands) on the Jeker and Meerssen (The Netherlands) on the Geul (see Figure 2.3). Their contributing areas are (in order) 2,245 km^2, 1,226 km^2, 430 km^2 and 338 km^2, respectively. The Membre station has a long discharge record dating from 1929. From 1929 to 1967, the daily values for Membre were based on one measurement per day, while after 1968 the daily values corresponded to the average values of hourly observations per day. The reason for selecting the other stations was the data availability for reconstruction of relatively long discharge records starting from the 1950s. The procedures of reconstruction and data quality control are described in section 2.3. These reconstructed discharge records together with the Borgharen and Monsin records have been used by Tu *et al.* (2004b and 2005a) for trend analysis and the trend results are included in this thesis.

Table 2.1 summarizes the origins of discharge/water-level data used in this study.

Table 2.1 Discharge and water-level data used.

River	Station	Record	Source
Meuse	Borgharen (N)[1]	Discharge (1911–2002)	RWS
	Monsin (B)[1]	Discharge (1911–2000)	
Rur	Stah (G)[1]	Discharge (1960–2003)	StUA
	Drie Bogen/Vlodrop (N)	Discharge (1891–1900)[2]	WRO
		Discharge (1953–1983)[2]	
Semois	Membre (B)[1]	Discharge (1929–2002)	MET-SETHY
Jeker	Jekermolenweg (N)	Water level (1952–1973)[2]	WRO
	Nekum (N)[1]	Discharge (1971–2001)	
Geul	Meerssen (N)[1]	Water level (1952–1973)[2]	WRO
		Discharge (1970–2001)	
	Gulpen (N)	Discharge (1972–2001)	
	Hommerich (N)	Discharge (1970–2001)	

Note: [1] The locations see Figure 2.2. [2] Data are available in hard copy. B = Belgium, G = Germany, and N = The Netherlands.

2.2.2 Precipitation data

Meuse basin upstream of Borgharen

Although there are many rain gauges installed in Belgium, only a few stations offer long and continuous daily records dating back to the beginning of the last century. Due to instrumentation, network and also (possibly) climatologically reasons, only observations after 1910 were found homogeneous and could therefore be used for studying the long-term evolution of precipitation (see Gellens, 2000). In this study, daily records (1911–2002) for seven Belgian gauging stations (see Figure 2.5) within the basin, provided by KMI, were used for statistical analysis. The period of record is as long as that for the discharge of the Meuse at Borgharen, allowing conclusions to be drawn regarding precipitation variability and rainfall-runoff relationship. These stations record precipitation amounts exceeding 0.1 mm/d at 8:00 AM (local time). In the upper part of the Meuse basin and its surroundings, ten French stations (see Figure 2.5) offer relatively complete daily records from 1928 to 1998, with the entire year of 1940 missing for all stations. Most French stations are geographically close to each other or to some Belgian stations. The French records from Météo-France were made available for this study later than the Belgian records.

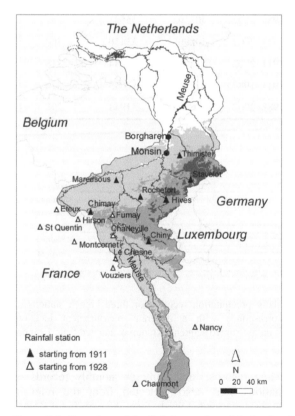

Figure 2.5 Overview of the Belgian stations (black triangles) and the French stations (white triangles) with long daily precipitation records in the Meuse basin and its surroundings.

The meta-data indicate that some of the station records were generated by combining short records from different locations and/or filled in using data from nearby stations. Table 2.2 provides information on the seven Belgian records. Figure 2.6 sketches the discontinuities of three French station records (Nancy, Chaumont and Vouziers) as examples. These stations were selected for estimation of the areal precipitation in the Meuse basin in section 2.4.2.

Table 2.2 List of seven Belgian stations with approximate coordinates.

No.	Station	Record	Latitude (d m s)	Longitude (d m s)	Elevation (m)
1	Denee Maredsous	1911–1981	501712	44603	222
	Mettet	1982–1991	501913	43925	238
	Saint-Gerard	1992–2002	502054	44406	240
2	Rochefort St.Remy	1911–2002	501034	51328	193
3	Thimister	1911–2002	503915	55148	266
4	Stavelot (P.C.)	1911–1988	502322	55534	300
	Stavelot	1989–2002	502303	55521	297
5	Hives	1911–2002	500905	53450	400
6	Chimay (Forges)	1911–2002	495852	42034	318
7	Chiny	1911–1987	494419	52045	370
	Lacuisine	1988–2002	494243	51943	298

Source: KMI.

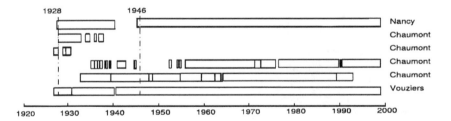

Figure 2.6 Schematic overview of daily precipitation records for three French stations as examples. Each line corresponds with a different measurement location. Continuous bars represent uninterrupted sequences (source: adapted from Leander and Buishand, 2004).

In addition to the above station records, long daily or monthly records of different lengths for a few stations were also collected from the relevant organisations (e.g. KMI and DWD) or downloaded freely from the website of KNMI (http://www.knmi.nl; new website http://eca.knmi.nl/). The stations include Aachen and Roermond in the lower part of the Meuse basin, Luxembourg Airport (Luxembourg) to the east of the basin, Ath, Uccle and Leopoldsburg (Belgium) to the west of the basin and De Bilt (The Netherlands) to the north of the basin. These station records have been used in synoptic-climatological analysis.

Geul and Rur subcatchments

Statistical analysis of precipitation was extended to the Geul and Rur subcatchments where the streamflows were also investigated in this study. No additional effort was made to collect the station records for the Semois and Jeker subcatchments. For the Geul subcatchment, seven Dutch and Belgian stations (see Figure 2.7) were used. For the Rur subcatchment, nine German and Dutch stations (see Figure 2.8) were used. Table 2.3 summarizes the precipitation data sources used.

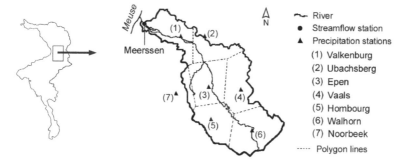

Figure 2.7 The precipitation stations in the Geul subcatchment, together with the Thiessen polygons used for calculating the areal precipitation (source: adapted from Agor, 2003).

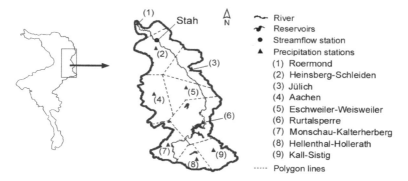

Figure 2.8 The precipitation stations in the Rur subcatchment, together with the Thiessen polygons used for calculating the areal precipitation (source: adapted from Calderón Hijuma, 2003).

2.2.3 Other meteorological data

Besides precipitation, evapotranspiration (ET) is another meteorological variable of interest in this study. The Penman-Monteith method is generally considered superior for estimating potential evapotranspiration (PET) of a reference grass crop (Jensen *et al.*, 1990). In this study, the PET of grass for the long period 1911–2002 has been computed for a Dutch station De Bilt (some 180 km north of Borgharen), for which

Table 2.3 Precipitation data sources for the Geul and the Rur subcatchments.

Subcatchment	Station[1]	Daily record	Source
Geul	(1), (2), (3), (4) and (7) (N)	1952–2001	KNMI
	(5) and (6) (B)	1973/1976–2001	KMI
Rur	(1) (N)	1869–2001	KNMI
	(2), (3), (5), (7), (8) and (9) (G)	1953–2002	DWD
	(4) (G)	1891–2002	DWD
	(6) (G)	1948–2002	WVER

Note: [1] Numbers in parentheses represent the stations indicated in Figures 2.6 and 2.7, respectively. B = Belgium, G = Germany, and N = The Netherlands.

the meteorological data are available for application of the Penman-Monteith equation. Daily data of relative sunshine duration, wind speed and temperature were freely available from the website of KNMI. The relative humidity data was specially requested. Meta-data for the wind speed record indicates that there was an instrument change after 1961 and different measurement heights were used for three periods 1911–1960, 1961–1990 and 1991–2002. Meta-data for the temperature record documented several minor changes in location, observation height or instrument. Meta-data for the records of relative sunshine duration and relative humidity were not available. As being situated in the same climatic region, the De Bilt station is expected to show a similar PET pattern to the meteorological stations located inside the Meuse basin. Preference should be given to the meteorological stations inside the basin, e.g. Maastricht Airport. However, no humidity data for this station was available. Moreover, the station was moved from the town to the airfield during the period of interest (1911–2002). In this study, the short PET record (1952–2001) for Maastricht computed based on the Makkink formula for grass, an alternative method for the Penman-Monteith equation and based on radiation and temperature data alone, was provided by KNMI. From 1952 to 1964, the global radiation was computed based on the relative sunshine duration, while from 1965 onwards the measured global radiation was used. The Makkink formula has been used for routine calculations of crop-reference ET by KNMI since 1987 (see De Bruin and Stricker, 2000).

Besides the above meteorological data used for estimating PET, daily temperature records for the meteorological stations Uccle and Luxembourg Airport in the surroundings of the Meuse basin were also collected from the website of KNMI. The temperature data are specially used for checking the weather conditions (frost) of winter floods in the Meuse.

2.3 Reconstruction of the discharge records

2.3.1 Procedure and methods

Initially, water levels at gauging stations were observed and recorded manually (e.g. two readings per day). Since the beginning of the 1970s, automatic water-level recorders (e.g. 15-minute-interval measurements, totally 96 readings per day) were introduced in the Meuse basin and data were then often stored in a digital form. Because many stations were moved or abandoned during the 1970s, most of the current stations on the tributaries have water-level or discharge records available from the late 1960s or later. In this study, the reconstruction of daily discharge records covering a period of almost 50 years has been carried out for the selected tributaries: the Geul, the Jeker and the Rur.

The procedure of reconstruction generally includes the following steps:

i) digitising the hard-copy data of water level (and discharge);

ii) converting the regular water-level measurements to discharges using the suitable rating curves (based on the formula in Appendix B);

iii) screening of suspicious values in both the hard-copy based discharge records and the digitised records;

iv) evaluating the influence of different observation frequencies (e.g. two readings per day or 96 readings per day) on the daily mean discharges;

v) correcting the evident non-homogeneous subsets by the double mass technique or adjusting daily values based on the regression equations;

vi) filling in the missing values by means of linear interpolation, linear regression on a monthly basis using the observation data of nearby stations which show strong correlation, or even by modelling; and

vii) combining the short records into the relatively long records for the stations of interest and checking the consistency of the reconstructed record by the double mass technique.

The principle of the double mass technique and the procedure of correction are given in Appendix A.

2.3.2 Geul river

The outlet of the Geul river is near Meerssen, some kilometres north of Maastricht. Both hard-copy data of water levels (November 1952–1973) and digitised daily discharge data (1970–2001) were available for Meerssen. The manual gauging station is some 200 m downstream from where the automatical recorder is located. Using the two records, Agor (2003) has reconstructed a daily discharge record for Meerssen from 1953 to 2001. The twice-a-day discharges were determined from the manually observed water levels using a rating curve and the daily mean discharges were then computed. Consistency of the digitised discharge record and the hard-copy based discharge record were checked for the overlapping period (1970–1973). Although a good correlation (correlation coefficient $r > 0.90$) was found between the two records, the double mass plot suggests that the hard-copy based discharges are on average approximately 8% less than the digitised discharges. The available digitised records for Hommerich (some 15 km upstream of Meerssen) on the Geul and Gulpen on the main tributary Gulp were used to identify which data set is more reliable. For the common period, the correlation results suggest that the digitised data for Meerssen appears more accurate. Therefore, taking the digitised data for Meerssen as a reference record for the Geul, the hard-copy based discharges from 1953 to 1969 were adjusted based on the relations between the two records over the common period. Figure 2.9 shows the scatter plot of the adjusted hard-copy based discharges against the digitised discharges. The adjusted hard-copy data is regarded as generally consistent with the digitised data obtained after 1970. For the entire reconstructed record (1953–2001), more than 800 missing values (including suspicious values), accounting for about 4.8% of the total daily values, were filled in by regression analysis using the data of the nearby stations (Hommerich, Gulpen and

Nekum in the order of preference) or using the rainfall-runoff simulation results (only for a short period from 1 January 1963 to 3 March 1963). Since the data from Hommerich are essential for filling in the missing values (nearly 700 values) at Meerssen, correction of non-homogeneity in the record for Hommerich has been made using the double mass technique.

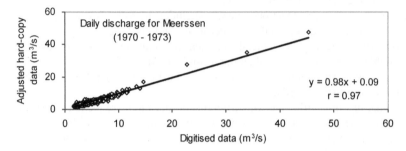

Figure 2.9 Scatter plot of the adjusted hard-copy based discharges and the digitised discharges for Meerssen (source: reproduced from Agor, 2003).

2.3.3 Jeker river

The Jeker or Geer river enters into the Meuse at Maastricht after crossing the Albert Canal. Most of its course is in Belgium. The last 5 km are located in The Netherlands (the Dutch Province of Limburg) where its name is written as Jeker. Two gauging stations are operated on the downstream Jeker: one is the Dutch station Nekum; the other one is the Belgian station Kanne, a few kilometres upstream of Nekum. In this study, digitised discharge data for Nekum from (February) 1971 to 2001 and hard-copy data of water level for the nearby downstream site Jekermolenweg from (November) 1952 to (September) 1973 were available. Discharge data for Kanne from 1975 to 1999 was only used for checking the reliability of the data at Nekum and completing some missing values in the record. According to WL (1988), the observation record at Nekum was considered as reliable. Therefore, in this study the Nekum station was taken as the reference station for the Jeker.

Double mass analysis of the discharge data for Nekum and Kanne indicates an inconsistency around April 1991 over the concurrent period 1975–1999. The mean value before April 1991 for Nekum was nearly 11% higher than that for Kanne, while the mean value after April 1991 for Nekum was about 7% lower than that for Kanne. Comparison was further made using the daily record for Meerssen on the Geul river, which has been carefully checked by Agor (2003), and the data from an adjacent tributary Geleenbeek. The double mass plots suggest that the Kanne record appears more accurate. Correction of the second subset (May 1991–1999) for Nekum was therefore made using the double mass technique. Using the same correction equation, the record of the latter period (2000–2001) was also adjusted. The water-level data in hard copy for Jekermolenweg was converted into the discharge data using three rating curves valid for the different periods. Consistency of the hard-copy based discharge record for Jekermolenweg and the digitised record for Nekum was checked for the overlapping period from February 1971 to September 1973. An inconsistency around May 1972 was observed. Good correlation ($r \geq 0.90$) was found for their second subsets after May 1972. The first subset before May 1972 for Nekum appears unstable. Therefore, it was decided that the hard-copy based discharge record for Jekermolenweg before May 1972 was used in combination with the digitised discharge record for Nekum after May 1972. Based on the linear relationship between the second subsets of the two records, the hard-copy based discharges at Jekermolenweg were adjusted with emphasis on the high flows. The consistency of the adjusted record for Jekermolenweg was checked using the record for Meerssen by plotting the double mass curve. The linear relationship has obviously improved. For the entire reconstructed record of daily discharge (1953–2001), totally more than 800 days of missing values and suspicious values have been completed by linear interpolation or by regression on a monthly basis using the data from Kanne and Meerssen.

2.3.4 Rur river

The Rur river enters into the Meuse at the Dutch town of Roermond. Its name is written as Roer when the river reaches the Dutch territory at Vlodrop. There are two gauging stations on the downstream part of the Rur. One is the German station Stah, which is close to the border with The Netherlands. The other one is the Dutch station Drie Bogen (also known as Vlodrop, with a station move), approximately 5 km downstream of Stah. The Stah station is of interest in this study.

A complete daily discharge record from 1960 to 2001 was available for Stah. One unpublished Dutch report (RWS, 1984) provided the daily discharge data for the Rur from 1953 to 1983 without missing values. According to this report, the data from Drie Bogen agree well with those from Stah for the period analysed. Therefore, in this study it was decided to directly use the data in the report to generate a record from 1953 to 1960 for Stah. First, the correlation of the discharges at Stah and the discharges at Drie Bogen as given in the report was obtained per month ($r > 0.90$ for all months except for September, and $r = 0.84$ for September) for the longest common period 1960–1983, with the data for Drie Bogen being shifted one day ahead. Next, the linear regression equations obtained for each month were used to derive the data for Stah from the data at Drie Bogen. Reconstruction of the daily record for Stah (1953–2001) was carried out by Calderón Hijuma (2003).

2.4 Estimation of the areal precipitation

2.4.1 Data completion and consistency of the station records

Meuse basin upstream of Borgharen

The collected precipitation records of the Belgian and French stations may suffer from inconsistency due to the effect of combining shorter records at different locations or filling in a large amount of missing values. Pronounced location and instrument changes usually result in shifts, instead of trends, in the precipitation data series. Prior to using these observation data, such potential inconsistencies in the station records should be carefully examined.

The station records for the entire period were examined independently (without a reference series) based on the annual totals (calendar years), which are less correlated and can be assumed independent. The missing data of the entire year of 1940 for each French station were filled in on a monthly basis, simply by taking the monthly average value of the data over the period 1928–1998 for each month excluding 1940. The station series were visually inspected first and then tested using the change point tests noted in section 2.8.2. The record for Hirson (geographically close to Chimay) was found suspicious and, thus, not considered in further change point analysis. The outcomes of the statistical tests are summarized in Table 2.4. Evaluation of the change point results was mainly based on the results given by the Pettitt test (Pettitt, 1979; see Appendix A), for which the first year after a change was used in the following text and the first year before a change was used in the following table.

Table 2.4 Change point results for the Belgian and French station records based on the annual precipitation totals (calendar years).

No.	Station record	Serial correlation r_1[1]	Pettitt test First year before change[2]	SNHT test First year after change[3]
	The Belgian station records (1911–2002)			
1	Maredsous	**0.265**	**1938 (0.97)**[4]	**1934 (11.99)**
2	Rochefort	0.149	1979 (0.77)	1912 (4.92)
3	Thimister	**0.286**	1949 (0.67)[4]	1997 (5.30)
4	Stavelot	0.100	**1936 (0.80)**	1997 (7.43)
5	Hives	**0.273**	**1952 (0.81)**[4]	1952 (6.77)
6	Chimay	**0.215**	**1947 (0.99)**[4]	**1978 (13.49)**
7	Chiny	0.012	1931 (0.50)	1999 (2.95)
	The French station records (1928–1998)			
8	Charleville	0.164	**1949 (0.89)**	**1935 (9.48)**
9	Chaumont	-0.101	**1976 (0.94)**	1976 (6.57)
10	Le Chesne	0.055	**1978 (0.83)**	1978 (4.24)
11	Eteux	0.161	**1964 (0.97)**	1978 (7.52)
12	Fumay	0.160	**1973 (0.98)**	**1973 (8.33)**
13	Montcornet	-0.111	1978 (0.67)	1978 (2.76)
14	Nancy	-0.103	**1964 (0.95)**	1964 (5.40)
15	Quentin	0.072	**1976 (0.97)**	**1978 (7.91)**
16	Vouziers	-0.136	**1977 (0.81)**	1932 (3.30)

Note: Probabilities for the Pettitt test and test statistics for the SNHT test are given in brackets. r_1 = serial correlation coefficient for lag 1. The bold values are significant: [1] at a significant level of 5%; [2] at a probability level of 80%; [3] at a significance level of 90%. [4] after pre-whitening the data series.

The split-record test results confirm the significance of the change points for most stations (except for Vouziers). Although statistically significant change points were found for the Belgian stations Maredsous (1939), Stavelot (1937), Hives (1953) and Chimay (1948), they do not correspond with their discontinuities due to station move indicated in Table 2.2 or can not be simply explained by the record data itself. No change point was identified for Chiny despite its station move in

1988. The majority of the French station records also show significant change points. Attention was paid to the records for Chaumont, Nancy and Vouziers, since these stations were selected to represent the upper part of the Meuse basin in estimation of the areal precipitation over the period 1928–1998. The station records for Chaumont and Nancy show significant change points around 1977 and 1965, respectively, which also do not correspond with their discontinuities due to station move or record gap indicated in Figure 2.6.

During the 20th century, the methodology for precipitation measurement has slightly changed. In 1951, the Belgian precipitation network was re-organised and a new type of rain gauge was introduced (Leander and Buishand, 2004). However, it was assumed that these instrumental changes did not have a major impact on the change point results of the station records. Since the change point years identified in the selected station series do not correspond with their records' discontinuities rising from location changes and also no other meta-data are available to indicate these changes, the station records can not be arbitrarily discarded. Taking climate variability into consideration, one would expect similar changes in the different station records in the same climatological region. A reason for the varying change points in the station records could be due to the orographic effects or random property of the data itself. At this stage, the seven Belgian station records (1911–2002) and the selected three French station records (1928–1998) are regarded as generally reliable.

Geul and Rur subcatchments

Data completion and quality control of the station records for the Geul and Rur subcatchments were performed by Agor (2003) and Calderón Hijuma (2003), respectively. Missing data of a few station records were interpolated through linear regression ($r > 0.90$), using the data from the neighbouring stations that show significant correlation in a spatial context. The function used to define the spatial correlation of two rain-gauging stations is given in Appendix B.

After data completion, the double mass technique was applied to check the quality of each station record by comparing the accumulated annual precipitation values of one station with those of the neighbouring stations in the subcatchment over the common period. No obvious errors were found on the annual basis. The station records used for both catchments can be used for climatic analyses.

2.4.2 Meuse basin upstream of Borgharen

Estimation of precipitation over an area is an important aspect of hydrological analysis. The simplest method of obtaining the average depth of precipitation over a basin is to average arithmetically the gauged amounts in the area. Theoretically, this method is appropriate only where precipitation gauges are randomly distributed and the terrain is uniform so that the individual gauge catches do not vary widely from the mean. These limitations can be partially overcome if topographic influences and areal representativity are considered in the selection of gauge sites (Linsley *et al.*, 1982). In this study, the arithmetic mean method was applied to estimate the areal precipitation in the Meuse basin upstream of Borgharen. Due to the different record lengths, two sets of stations were used:

i) seven Belgian stations with the daily records from 1911 to 2002; and

ii) three French stations with the monthly records from 1928 to 1998, together with seven Belgian stations initially selected.

As seen in Figure 2.5, the seven Belgian stations are almost uniformly distributed in the central part of the Meuse basin. Inclusion of a few French stations in the upper part of the basin is expected to provide a better representation of the areal precipitation for the area upstream of Borgharen. Three French stations Chaumont, Nancy and Vouziers were therefore selected. In this study, the areal precipitation record obtained from the first set of seven stations is referred to the "MeuseLP" record, while the areal precipitation record obtained from the second set of ten stations is referred to the "MeuseSP" record. Due to the averaging, the areal precipitation series are more likely to exhibit less variability than the individual station series. In a very late stage of this study, for the HBV modelling of the Meuse basin, Ashagrie (2005) derived a new areal precipitation record from 1911 to 2000 (referred to as the "MeuseRLP" record in this study) through the area weighing method from the extended areal precipitation records of the 15 subcatchments upstream of Borgharen (see section 6.5.2).

Table 2.5 gives the basic statistics of the three areal precipitation records for comparison. It appears that the MeuseLP record probably "overestimates" the areal precipitation in the Meuse basin. Over the common period 1912–2000, the mean annual value of the MeuseLP record is about 80 mm higher than that of the MeuseRLP record. Despite the slight difference in the precipitation amounts, strong correlation ($r > 0.99$) was found between the two records on the daily basis. The mean values of the MeuseSP record are closer to those of the MeuseRLP record.

Table 2.5 Mean values (mm/a) of annual, winter and summer precipitation totals in the Meuse basin based on the different records.

Record	Period	Annual (Nov–Oct)	Winter (Nov–Apr)	Summer (May–Oct)
MeuseLP	1912–2002	1034	528	507
	1912–2000	1031	524	507
	1929–1998	1032	521	511
MeuseSP	1929–1998	958	479	479
MeuseRLP	1912–2000	949	490	459

In this study, including the previous relevant publications by Tu *et al.* (2004a, 2004b, 2005a, 2005b, 2005c and 2006), the MeuseLP record has been used for trend analysis of precipitation in the Meuse basin. The limited number of gauging stations used is obviously far from adequate for an accurate estimation of the areal precipitation over the area. However, this study emphasises the temporal variability of precipitation. It was assumed that the climate-induced precipitation change in that area, if obvious, was large enough to be detected from the MeuseLP record. Besides, the potential influence of discontinuities of the station records will be also considered. For the water-balance analysis, the accuracy of the areal precipitation estimates is required. Therefore, the MeuseRLP record has been used in this part.

2.4.3 Selected subcatchments (Geul and Rur)

The Thiessen polygon method was applied to estimate the areal precipitation for the Geul and Rur subcatchments (see Agor, 2003; Calderón Hijuma, 2003). The Thiessen polygon lines are shown in Figures 2.6 and 2.7, respectively. For the Rur subcatchment, nine German and Dutch station records from 1953 to 2001 were used. For the Geul subcatchment, two sets of station records have been used:

i) five Dutch station records from 1952 to 2001; and

ii) five Dutch station records from 1952 to 1976, and five Dutch station records together with two Belgain station records from 1977 to 2001.

Accordingly, the derived daily records of areal precipitation for the Geul subcatchment are referred to the "Geul5P" and the "Geul7P", respectively. In the study by Agor (2003), the Geul5P record was used for trend analysis, while the Geul7P record was used only for the water balance analysis and the rainfall-runoff modelling study. For the common period 1977–2001, the mean annual value (908 mm/a) of the Geul5P record was found about 16 mm (less than 2%) lower than the mean annual value of the Geul7P record.

The areal precipitation data derived for the two subcatchments have been compared with the areal precipitation data in the Meuse basin over the common period 1954–2001. Good correlation was obtained on the monthly basis ($0.81 \leq r \leq 0.95$). The mean annual values for the Geul (891 mm/a, based on the Geul5P record) and the Rur (836 mm/a) catchments are about 150–200 mm lower than the mean annual value for the Meuse basin (based on the MeuseLP record).

2.5 Estimation of potential evapotranspiration

2.5.1 Penman-Monteith method

The Penman-Monteith formula for estimating potential evapotranspiration (PET) of grass with a length of 12 cm (see Appendix B) was used to estimate PET for De Bilt (further referred as PET_{P-M}). The quality of the relevant meteorological data, i.e. relative sunshine duration, wind speed, temperature and relative humidity, has been checked carefully through change point analysis of their annual mean daily values (calendar years). After corrections made due to different observation levels of the wind speed data (to a height of 2 m for the entire study period), no obvious change was found in the annual mean values. The record of annual mean daily temperature shows an obvious increase after 1989, which, however, does not coincide with the measurement changes according to the available meta-data. Statistically significant increases were found in both the annual mean daily relative sunshine duration and the annual mean daily relative humidity, with the change points around 1989 and 1950, respectively. In the absence of meta-data, no correction has been made to both data series. The meteorological data noted above can be regarded as generally suitable for computing PET_{P-M}. A daily PET_{P-M} record for De Bilt from 1911 to 2002 (ranging from 0.01 mm/d to 7.51 mm/d) has been prepared and used for analysis of ET fluctuation in the Meuse basin. The PET_{P-M} record yields a mean annual total of 563 mm/a. Correlating the annual PET_{P-M} totals with the annual mean values of the four parameters indicates that relative sunshine duration and relative humidity are the dominating parameters, their r values being above 0.70. The correlation with the annual mean daily wind speed is the poorest ($r = 0.11$, not statistically significant). The correlation with the annual mean daily temperature is intermediate ($r = 0.46$).

2.5.2 Makkink formula

The Makkink formula for grass (see Appendix B) has also been applied to compute PET for De Bilt (further referred as $PET_{Makkink}$) from 1911 to 2002. The $PET_{Makkink}$

record gives a mean annual total of 559 mm/a, about 4 mm lower than the mean annual total of the PET_{P-M} record. For the concurrent period 1953–2001, the mean annual total of the $PET_{Makkink}$ record for De Bilt (555 mm/a) is only 8 mm lower than that for Maastricht Airport (563 mm/a). Application of different formulas and use of the meteorological data of De Bilt in computing PET bring about only very small differences. Therefore, the derived PET_{P-M} record for De Bilt is suitable for analysis of ET fluctuation in the Meuse basin. The $PET_{Makkink}$ records for De Bilt and Maastricht Airport were not involved in the further analysis.

2.6 Synoptic data

2.6.1 Grosswetterlagen system

The *Grosswetterlagen* (European atmospheric circulation patterns) system according to Hess and Brezowsky (see Gerstengarbe and Werner, 1999) is a well-known subjective classification system in synoptic climatology, which describes the circulation patterns over Europe and the eastern part of the North Atlantic Ocean. The classification system recognises three groups of circulations divided into ten major types, 29 sub-types and one additional sub-type for the undetermined cases (Table 2.6). The three circulation groups are defined as zonal, half-meridional (mixed) and meridional, respectively. Bárdossy and Caspary (1990) briefly summarized the characteristics of major circulation patterns. For a detailed description of the classification, see Gerstengarbe and Werner (1999). A circulation pattern generally persists for several days while the entailed weather features remain constant. The zonal circulations, particularly the sub-type "West cyclonic" (Wz), are often associated with rain. The sub-types Southwest cyclonic (SWz) and Northwest cyclonic (NWz) within the half-meridional group refer to weather conditions similar to those associated with the sub-type Wz. Although additional information (500 hPa heights) was used in the classification since the late 1940s, no systematic non-homogeneity in the *Grosswetterlagen* record was found for the period 1930–1960, except that the major type SW showed a possible change point in the late 1940s (Bárdossy and Caspary, 1990).

Revised data series of the *Grosswetterlagen* system are available since 1881. The daily record from 1911 to 2002, collected from DWD, was used in this study to describe the relationship of the changing precipitation pattern in the Meuse basin to synoptic climatology. The missing values of a few dates (i.e. 27 March 1997, 6 March 2001 and 17 December 2002) were assigned to be the undetermined cases.

2.6.2 North Atlantic Oscillation (NAO) index

The use of circulation indices is commonly associated with synoptic climate studies. The most well-known circulation index over Europe is the North Atlantic Oscillation (NAO) index, which is associated with changes in the surface westerlies across the North Atlantic into Europe. The NAO index is traditionally defined as the normalised pressure difference between a station at the Azores and one at Iceland. Although the NAO occurs in all seasons, it is during winter that it is particularly dominant. During a high NAO index, the winter precipitation in Northern Europe may increase, while Southern Europe may be drier (Hurrell, 1995).

In this study, the monthly average values of the NAO index, defined as the

Table 2.6 Classification of the *Grosswetterlagen* system according to Hess and Brezowsky

Group of circulation types			
Major types		Sub-types	Abbreviation
		Zonal	
West	W	1. West anticyclonic	Wa
		2. West cyclonic	Wz
		3. Southern West	Ws
		4. Angleformed West	Ww
		Half-Meridional	
Southwest	SW	5. Southwest anticyclonic	SWa
		6. Southwest cyclonic	SWz
Northwest	NW	7. Northwest anticyclonic	NWa
		8. Northwest cyclonic	NWz
Central European high	HM	9. Central European high	HM
		10. Central European ridge	BM
Central European low	TM	11. Central European low	TM
		Meridional	
North	N	12. North anticyclonic	Na
		13. North cyclonic	Nz
		14. North, Iceland high, anticyclonic	HNa
		15. North, Iceland high, cyclonic	HNz
		16. British Islands high	HB
		17. Central European Trough	TrM
Northeast	NE	18. Northeast anticyclonic	NEa
		19. Northeast cyclonic	NEz
East	E	20. Fennoscandian high anticyclonic	HFa
		21. Fennoscandian high cyclonic	HFz
		22. Norweigian Sea-Fennoscandian high anticyclonic	HNFa
		23. Norweigian Sea-Fennoscandian high cyclonic	HNFz
Southeast	SE	24. Southeast anticyclonic	SEa
		25. Southeast cyclonic	SEz
South	S	26. South anticyclonic	Sa
		27. South cyclonic	Sz
		28. British Islands low	TB
		29. Western Europe trough	TrW

Source: Bárdossy and Caspary (1990). There is one additional sub-type (Ue) for the undetermined cases.

difference between the normalised sea level pressure in Gibraltar and the normalized sea level pressure in southwest Iceland (Reykjavik), were downloaded from the website of CRU (http://www.cru.uea.ac.uk/ftpdata/nao.dat). This Gibraltar-minus-Iceland version of the NAO index was originally developed by Jones *et al.* (1997) and later modified. It is suggested that the index should only apply to the winter half when using monthly values of the NAO index. A record from 1911 to 2002 was used in this study.

2.7 Observed changes resulting from human activities

2.7.1 Introduction

Definitions

The terms "land cover" and "land use" may be confusing. The two terms are sometimes used as if they are synonymous but they are not. Land cover refers to the physical characteristic of the earth's surface, captured in the distribution of vegetation, water, desert, ice, and other physical features of the land including those created solely by human activities such as mine exposures and settlement. Land use is the intended employment and management strategy placed on land cover type by human agents or land managers. Changes in land cover/use may be grouped into two broad categories: conversion or modification. Conversion refers to changes from one cover or use type to another. Modification refers to subtle changes that affect the attributes of the land cover without changing its overall classification (Baulies and Szejwach, 1997).

For a global change perspective, land cover is the more important property. Comparatively conversion from one land cover category to another is often well documented, whereas modification within one category is usually more difficult to "observe" and often ignored in large-scale investigations. Recently, there has been increased recognition of the importance of the modification of land attributes (Lambin *et al.*, 2003).

Historical land cover changes and recent rapid land use changes

The natural landscape in the Meuse basin has been fragmented and modified in the past centuries, especially through agricultural and forestry developments. Figure 2.10 depicts four striking stages in the evolution of the Ardenne landscape using the example of Noville.

Figure 2.11 presents the natural land cover patterns about 1,000 years ago and the current land use patterns (based on the recent CORINE data set) at the basin scale for comparison. Most notable changes in the basin are observed in the broad categories of forests, agricultural land (arable land and pastures) and built-up area. Currently, about 34% of the Meuse basin upstream of the Belgian-Dutch border is developed as arable land, 20% as pasture, 35% as forest, and 9% as built-up area (De Wit *et al.*, 2001).

It should be realised that the overall percentages of broad land cover categories are not sufficient to illustrate the land use changes. Over the recent decades, the most important change appears to be the intensification of land use through better management of production factors (Fresco, 1994). Total agricultural production in

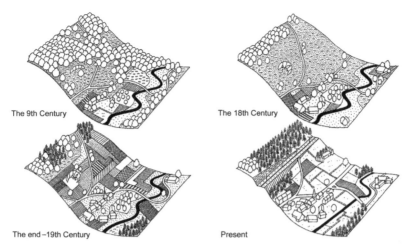

Figure 2.10 Evolution of the landscape in the Belgian Ardennes (source: DGRNE, 1996).

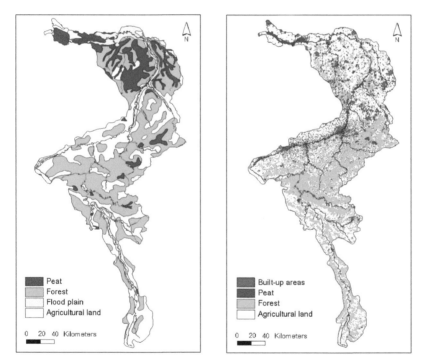

Figure 2.11 Comparison of historical (about 1,000 year ago) land cover patterns (left; source: adapted from De Mars *et al.*, 2000) and current land use patterns (right, source: based on the CORINE land cover database provided by RIZA) in the Meuse basin. The legends of both maps are highly simplified. Note that flood plain is not a separate class in the CORINE land cover database.

western Europe has increased significantly since the 1950s, to a great extent as a result of the introduction of high yielding crop varieties and changes in agricultural practice and techniques (DGRNE, 2000).

Scope of land use inventory and data sources

The inventory of land use changes in the Meuse basin refers to the French and Belgian parts. The French part includes parts of three departments Vosges (5,874 km²), Meuse (6,211 km²) and Ardennes (5,229 km²), see Figure 2.12. The Belgian part largely lies in the Walloon Region (also called Wallonia, about 16,845 km²), see Figure 2.13. The Walloon Region covers most part of the Ardennes Meuse.

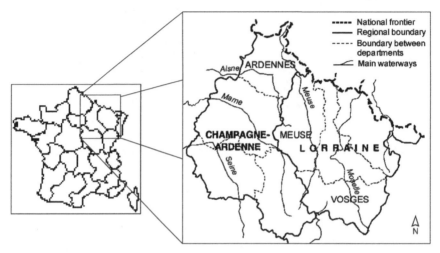

Figure 2.12 Location of the French departments Ardennes (in the Champagne-Ardenne Region), Meuse and Vosges (in the Lorraine Region).

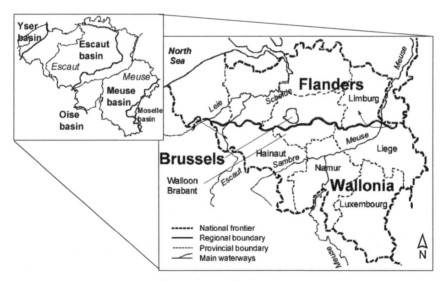

Figure 2.13 Location of the Walloon Region in Belgium.

Aerial photographs and satellite images for study of land use changes are generally available since the 1950s or later. The CORINE data set is based on satellite images as the primary information source and is the only data set providing a broad overview of land cover/use at European level enabling cross border investigations and comparisons at European level. This first CORINE land cover inventory was performed between 1986 and 1995 (EEA, 1995). To chart historical changes in land cover/use in the Meuse basin occurring in the 20th century, a variety of data sources have been searched, including official statistical yearbooks, regional statistical, scientific publications and project reports. The collection of relevant materials was indeed labour-intensive. Description of land use/cover changes in this study is mainly based on the readily-available data and information from the forementioned types of sources. No aggregation or grouping of these data was made due to differences in the categories of land use statistics and sometimes due to different periods used. This study emphasises the global evolution (long-term trend) of land cover/use changes in the basin in order to interpret the identified changes in the discharge regime of the Meuse river. Emphasis was given to the principal categories of agriculture area, forest area and urban area, and changes in agricultural practices. Information on wetland and land drainage in the area is poor and thus not included in the description. Since the Meuse river has in the past undergone many modifications particularly for navigation, hydraulic activities carried out in the Meuse river are included. Tu *et al.* (2005a) has provided a summary of land use changes and hydraulic activities in the Meuse basin.

A few tributaries of the Meuse river have been selected for case studies of streamflow. Two of them are located downstream of Borgharen, i.e. the Rur and the Geul. A brief description of the selected subcatchments are separately presented based on limited data and local information.

2.7.2 Changes in the forest area

The current percentages of forests and woodlands in the French departments Vosges, Meuse and Ardennes are about 49%, 37% and 29%, respectively (source: estimated based on the data taken from http://www.insee.fr and http://draf.champagne-ardenne.agriculture.gouv.fr). In the Walloon Region, the current percentage of forest area is about 32%, but with great variation within the region. The higher percentages are observed in the natural regions Ardennes (52%), Fagne-Famenne (44%) and Gaume (42%) to the south of the Sambre-Meuse Valley (DGRNE, 2003).

Historical data from local sources indicate that, over the past hundred years, the area of forests and woodlands in the French and Belgian parts has been rather stable and has even shown a slight increase (see Figure 2.14). Nevertheless, the forest types and management have experienced notable changes (mainly in the Belgian part). For example, the percentage of the conifer forest in the forest area of the Walloon Region has increased from 14% in 1895 to 50% in 1984, while the percentage of the deciduous broad-leaved forest has dropped at the same time. The current percentages of the coniferous and deciduous forests are 48% and 52%, respectively. Close to 90% of the region's forest area is managed for timber production (DGRNE, 2000). The highest proportion of conifers is found in the (Belgian) Ardennes region, which is largely explained by economic reasons in timber production such as greater productivity and shorter production cycles.

Over the last few decades, northwestern Europe has witnessed the intensification of wood-producing techniques and practices such as mechanisation of the planting and harvesting of trees, the draining of "wet" forests etc. (EEA, 1995). The forestry infrastructure and networks of roads and paths for tourism can considerably influence the drainage network in the forest areas.

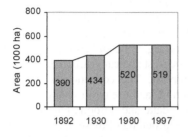 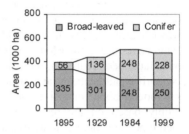

Figure 2.14 Changes in the forest (and woodland) areas in the French departments Vosges and Meuse (left) (source: the bar diagram is made based on the data provided by SCEES) and in the Belgian Walloon Region (right) (source: reproduced from DGRNE, 2000).

2.7.3 Changes in the agricultural area

The French part of the Meuse basin is largely rural. The utilised agricultural area in the Walloon Region accounts for 45% of the total area (in 2002), nearly half being grassland. The high proportion of arable land is found in and to the north of the Sambre-Meuse Valley, where the soils are fertile and the land is predominantly used for cereal, sugar beets and potatoes. In the south of the valley, a large part of the utilised agricultural area is used for meadows/pastures and fodder crops (DGRNE, 2003).

A long sequence of the utilised agricultural area data was available for the French departments Vosges and Meuse. Figure 2.15 depicts the long-term trend. Over the past century, the utilised agricultural area appears to be slightly decreasing till the early 1970s and after which it remains relatively stable. A more evident change is found in the proportions of arable land and permanent pasture. After the area of permanent pasture exhibits a distinct increasing trend till the early 1970s, it has been decreasing until the present. The arable area showed the opposite change pattern. According to DGRNE (1993, 2000 and 2003), the utilised agricultural area in the Walloon Region slightly decreased between the 1950s and the 1970s and has remained relatively stable since 1980. A detailed analysis further indicates a subtle decrease until 1992. In the last two decades, the provinces of Liège and Luxemburg show a pronounced reduction of the utilised agricultural area. The area of meadows and pastures in the region has obviously decreased between 1977 and 1992, with the most pronounced change in and to the north of the Sambre-Meuse Valley, and it appears to stabilise since 1995. The area for the main crop of cereals has been decreasing since the beginning of the 1980s, currently accounting for 24% of the utilised agricultural area (in 2002). Sugar-beet is the main industrial crop cultivated in the region, accounting for about 8%. Its area has increased between 1977 and 1981 and after which it decreased.

2.7.4 Changes in the urban area

After a rapid increase in the 19th century and a slower growth rate in the first half of the 20th century, the western European countries were highly urbanised by 1950. Nowadays, Belgium is one of the world's most heavily urbanised nations, with more than 95% of the population living in urban areas. Industrial activities are concentrated along the Sambre downstream from Charleroi and along the Meuse

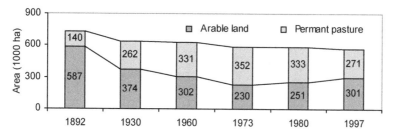

Figure 2.15 Change in the utilised agricultural area (arable land and permanent pasture) in the French departments Vosges and Meuse (source: the bar diagram is made based on the data provided by SCEES, 2003).

upstream and downstream from Liège (DGRNE, 2000). A few important urban centres in the French part are Verdun, Sedan and Charleville-Mézières.

Urbanisation is generally accompanied by a reduction of agricultural land. The period 1950–1960 is often seen as a turning point regarding the industrialisation of farming, considerable urban development and the expansion of transport infrastructure (DGRNE, 2000). It was estimated that, between the 1950s and the 1980s, the built-up area in many European countries has increased by 25 to 75% (EEA, 1995). In the Walloon Region, the urban areas accounted for 10.7% in 1980, 12.7% in 1996, and presently for more than 13.5%. The increase in built-up area and the decrease in agricultural land were particularly marked in the north of the Sambre-Meuse Valley. While it appears that most of the agricultural land has been converted to the urban area, it is nevertheless difficult to identify the exact transfers between the land use categories, as the processes may be indirect, i.e. involving intermediate transfers towards other types of land use (DGRNE, 2000 and 2003).

The steady urbanisation, dispersed housing pattern and dispersal of economic activities usually bring about the simultaneous increase in infrastructure and road construction and increased mobility requirements.

2.7.5 Changes in land use management

Intensification of agricultural production is linked to mechanisation and motorisation of farming, and modifications in crop rotation systems. The process can be illustrated by the changes in, for example, the number and size of farms and the use of machinery. Generally larger farm sizes are more efficient in terms of economic costs and inputs of labour. However, the use of tractors (particularly large and heavy vehicles) can lead to compaction of some types of soils. In the Walloon Region, the number of horticultural and agricultural farms has obviously decreased since 1980, mainly as a result of the transition towards industrial crops. Meanwhile, the average size of the farms has grown, particularly in the areas with large-scale cultivated crops (DGRNE, 2000). Figure 2.16 shows the number of tractors in use for agriculture in Belgium-Luxembourg and The Netherlands. There is an obvious increasing trend from the 1960s to the mid-1980s, followed by a slightly decreasing trend.

The irrigated areas are mainly limited to the most important vegetable growing areas and the sugar-beet growing areas, which are most notable in the Sambre-Meuse Valley. The irrigation area in the Walloon Region (in 1997) accounts for only 0.7% of the utilised agricultural area (DGRNE, 2000).

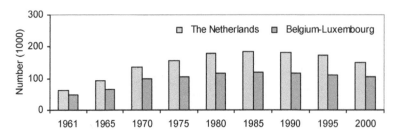

Figure 2.16 Number of tractors in use for agriculture (1961–2000) in Belgium-Luxembourg
and The Netherlands (source: FAOSTAT Data base, http://apps.fao.org).

2.7.6 Changes in the river network

Reservoirs

Some river valleys in the Ardennes and Eifel mountains have been dammed to
create reservoirs. The largest reservoirs are located in the upper branches of the
Viroin, Semois, Sambre, Amblève, Ourthe, Vesdre (see Figure 2.17) and Rur (see
Figure 2.21). These reservoirs are used for various purposes such as hydroelectricity
production, drinking and industrial water supply, flood control and (low) flow
regulation. Limited information is available about the operation of these reservoirs.
The Eau d'Heure reservoir complex is in fact composed of five reservoirs. Its main
purpose is to regulate the flow of the Sambre (1–5 m³/s) for the Charleroi-Bruxelles
Canal during dry seasons. Its additional objectives include flow regulation for the
Meuse river and reduction of the effects of industrial and urban pollution on the
downstream river (Berger, 1992). Most of the reservoirs were built in the second
half of the 20th century. Figure 2.18 depicts the changes in the total capacity and the
total drainage area of the largest Belgian reservoirs over time. The combined
drainage area (including the nested drainage area of the Bütgenbach reservoir, but
excluding the drainage areas of the Coo reservoirs) covers some 1,420 km²,
accounting for about 6.7% of the basin area upstream of Borgharen. The combined
reservoir volume (including the Coo reservoirs) is about 8 million m³.

Weirs and canals

Shipping has played an important role in many hydraulic works in the Meuse since
the 19th century. Nowadays almost the entire stretch of the Ardennes Meuse is
completely regulated with weirs and thus is navigable. As a result, the water levels
and discharges in the river are frequently affected (Berger, 1992). In addition, there
are a number of canals fed by water of the Meuse. Downstream of Liège, the Albert
Canal (Dutch: Albertkanaal), the South-William Canal (Dutch: Zuid-Willemsvaart)
and the Juliana Canal (Dutch: Julianakanaal) are the most important ones (see Figure
2.4). These canals are not only used for navigation but also play a crucial role in the
water supply for Brussels, Flanders and the southern part of The Netherlands.

Alterations in the river channel

The navigable stretches of the Sambre and the Meuse have been dredged frequently
(DGRNE, 2003). Micha and Borlee (1989) made a detailed description of the

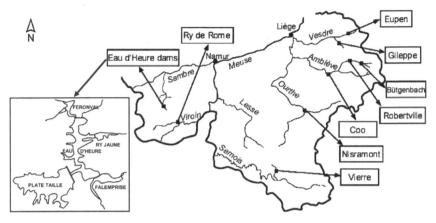

Figure 2.17 Locations of the relatively large reservoirs in the Belgian Walloon Region (source: adapted from Mergen, 2002; also see http://voies-hydrauliques.wallonie.be).

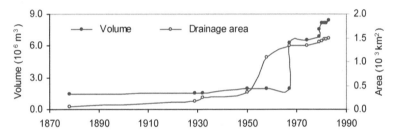

Figure 2.18 Total volume and total drainage area (including the nested drainage area of the Bütgenbach reservoir, but excluding the drainage areas of the Coo reservoirs in the Amblève) of the largest 14 reservoirs in the Walloon Region, Belgium. The graphs are made based on information from Mergen (2002).

alterations in the Walloon part of the Meuse channel since 1800. The first systematic alterations started in 1839 near the Namur part, transforming the deep stretches into navigatable passes. After the 1880 flood, it was proposed that the curves of the river be straightened and that the channel be normalised by regulating its slope and by fixing standards of minimum width. The process of modernization began from 1918, leading to the division of the Belgian Meuse into the lower Meuse (from the Dutch border to Namur) and the upper Meuse (from Namur to the French border). The lower Meuse was the first to be developed. A new program to improve navigation on the Meuse began around Liège in 1923 and was implemented in 1927 in response to the disastrous flooding of 1926. Since the 1960s, navigation of the Meuse has progressed. It was only in 1983 that the modernization program set up in 1960 passed Namur and continued upstream towards the French border. During the modernization period from 1918 until present, many hydraulic measures have been taken in the lower Meuse channel to remove water more rapidly from the area. The works include the removal of islands, straightening the natural banks, and deepening and widening certain sections of the river (Micha and Borlee, 1989).

Floodplains

The floodplain of the Meuse changes from south to north. In the southern part, the Meuse Lorraine valley has a floodplain (used mostly as meadows) of some kilometres width, which may get inundated even during an average flood (see Figure 2.19). In the central part between Charleville-Mézières and Liège, the Ardennes valley is narrow and steep and little floodplain area is available. The width of the floodplain (winterbed) in the northern part generally varies between 200 m and 2 km (De Wit *et al.*, 2002).

Figure 2.19 Floodplain in the French part (along the tributary Bar) of the Meuse basin during an average flood (source: De Wit *et al.*, 2003).

2.7.7 Changes in the selected subcatchments

Rur subcatchment

Most of the Rur subcatchment is located in Germany. The upstream part is hilly with poorly permeable rocks, whereas the downstream part is flat with permeable soils. The important tributaries include the Olef, Kall, Inde and Wurm. Currently, about one third of the German part is covered with forest, mainly distributed in the mountain area, and one third is used as arable land, mostly in the lowland area (source: WVER). Half of the remaining land is used for pastures and the other half for urban area. Figure 2.20 shows the current land use map. The important industrial cities in the German part are Aachen, Düren and Jülich.

In the 20th century, nine dams (Figure 2.21) have been constructed in the upstream part for water supply, flow regulation and hydroelectricity, holding a total volume of approximately 300 million m^3. The Urfttalsperre reservoir was built in 1905, which was the largest one in Europe at that time. Most of the other reservoirs were completed in the 1950s. Six reservoirs are operated on the basis of having the largest Rurtalsperre reservoir (see Figure 2.22), being the second largest dam of Germany, as the balancing reservoir. The smaller reservoir at Heimbach, which was built in 1935 with a capacity of 1.2 million m^3, controls the water release to the lower reach of the Rur. The area upstream of Heimbach sizes about 675 km^2.

In the triangle Aachen, Köln and Mönchengladbach, the largest lignite (often referred to as brown coal) resources of Europe are located. Lignite mining started in the 18th century and became mechanised by the end of the 19th century with a production of more than 10 million ton in 1910. Figure 2.21 indicates the locations of mines in the Rur subcatchment and its surroundings. The open pit mine at Inden, with a depth of 230 m and a surface area of 47 km^2, is in use since 1981. The mine

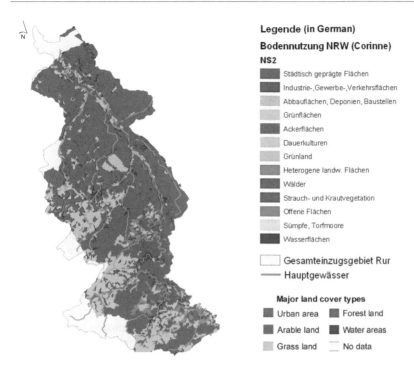

Legende (in German)

Bodennutzung NRW (Corinne)

NS2

- Städtisch geprägte Flächen
- Industrie-,Gewerbe-,Verkehrsflächen
- Abbauflächen, Deponien, Baustellen
- Grünflächen
- Ackerflächen
- Dauerkulturen
- Grünland
- Heterogene landw. Flächen
- Wälder
- Strauch- und Krautvegetation
- Offene Flächen
- Sümpfe, Torfmoore
- Wasserflächen

☐ Gesamteinzugsgebiet Rur
── Hauptgewässer

Major land cover types

- Urban area
- Arable land
- Grass land
- Forest land
- Water areas
- No data

Figure 2.20 Current land use in the German part of the Rur subcatchment (source: adapted from the map provided by WVER).

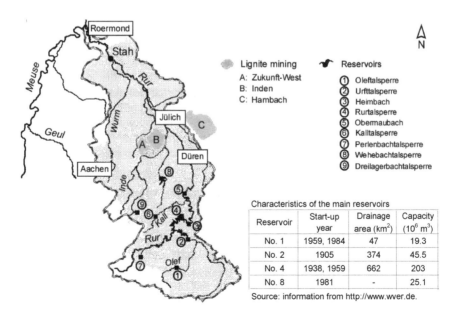

Lignite mining
A: Zukunft-West
B: Inden
C: Hambach

Reservoirs
① Oleftalsperre
② Urfttalsperre
③ Heimbach
④ Rurtalsperre
⑤ Obermaubach
⑥ Kalltalsperre
⑦ Perlenbachtalsperre
⑧ Wehebachtalsperre
⑨ Dreilagerbachtalsperre

Characteristics of the main reservoirs

Reservoir	Start-up year	Drainage area (km²)	Capacity (10^6 m³)
No. 1	1959, 1984	47	19.3
No. 2	1905	374	45.5
No. 4	1938, 1959	662	203
No. 8	1981	-	25.1

Source: information from http://www.wver.de.

Figure 2.21 Locations of the reservoirs and the lignite mining sites in the Rur subcatchment and its surroundings (source: adapted from Calderón Hijuma, 2003).

Figure 2.22 The Rurtalsperre reservoir on the Rur (source: http://www.wver.de).

at Zukunft-West, with a depth of 180 m and a surface area of 20 km², was in operation between 1950 and 1987 and is now closed. The mine at Hambach, just outside the subcatchment, came into operation in 1984 and extends to a maximum depth of 470 m, covering an area of 85 km². Around the same time, one mine at Garzweiler situated 15 km to the east of Hambach was opened. This mine covers an area of 56 km² and extends to a depth of 165 m. Serious impacts of the mining activities on the river regime in the area were felt from 1955 onwards, when the annual production was dramatically increased. The mining company pumps surplus water from Inden and some water from Hambach into the Rur river, but most of the surplus water of Hambach is discharged into the Erft, the river that runs to the east of the Rur (De Laat *et al.*, 2005; also see BUND, 2002). Groundwater flow in the Rur valley rift system is strongly influenced by the hydraulic properties of the faults. Isohypses for 1989 show hydraulic heads near Hambach of 180 m below mean sea level and at Inden around mean sea level or above (De Laat *et al.*, 2005). Mining activities in the German part of the Meuse basin in a long period have caused an extensive drawdown in the groundwater tables due to dewatering of mines or open pit areas. Stuurman and Vermeulen (2000) modelled the groundwater flow from the German border into The Netherlands and found a drop in the groundwater level by 30 m in some places of the deep aquifers. In the southern part of the neighbouring subcatchment Niers (1,350 km²), Meiners (2002) reported that about 80–150 million m³ groundwater per year has been extracted in a period of up to 40 years due to the lignite mining at Garzweiler II (some 15 km east of Roermond) and, consequently, the groundwater level has extensively lowered more than 200 m.

Semois subcatchment

The Semois flows into the Meuse near Monthermé in France. Most part of the subcatchment is located in Belgium (see Figure 2.23). The important tributaries are the Rulles and Vierre. The upstream course of the Semois lies in a relatively even area being part of the Belgian Lorraine. The strata in the upstream part have a varying permeability. After the mouth of the Vierre, the river flows through the Ardennes Massif with rocks of a palaeozoic origin. The gradient of that part of the river is greater than that in the even area. The forests account for about 40% of the area, 80% of which are deciduous forests (Berger, 1992). The agricultural area has reduced by several percent over recent two decades (DGRNE, 2003). There is a small dam (Barrage de la Vierre, see Figure 2.23) in the Vierre, which was built in 1967 (Mergen, 2002; 1970 given by Berger, 1992) for electricity generation.

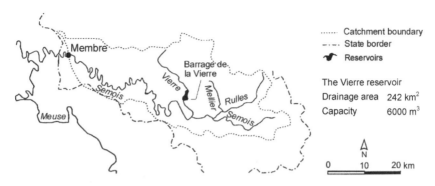

Figure 2.23 The map of the Semois subcatchment (source: adapted from Berger, 1992).

Jeker subcatchment

The geology in the Jeker subcatchment consists of the Cretaceous chalks with a varying thickness from a few meters in the south up to 100 m in the northeast. The groundwater table is located at depths ranging from 10 m to more than 40 m below the land surface. Most of the aquifer is unconfined, except in the north where semi-confined conditions prevail under the Jeker alluvial deposits. An important fault (the Horion-Hozémont fault), associated with a zone of highly fractured chalk, crosses the domain from southwest to northeast, greatly influencing hydrogeological conditions. The groundwater catchment extends beyond the surface catchment boundaries. The Jeker is the main outflow of the chalk aquifer. There has been groundwater abstraction from wells and drainage galleries in the area. Groundwater resources in the area provide drinking water of about 60 thousand m^3 per day for the city of Liège and its suburbs (Brouyère et al., 2004). The arable land dominates the land use in the subcatchment area.

Geul subcatchment

The Geul subcatchment is mostly situated in the Dutch Limburg Province. The subcatchment has a rolling topography. An extensive limestone plateau (containing faults and fissures in the bedrock) forms the middle part and the loess area occupies the biggest acreage. Groundwater abstraction in the area has been reported. The area has a history of industrial mining (occurred before the early 20th century).

Dautrebande et al. (2000) have made an inventory of land use changes in the area over the last 50 years for hydrological modelling of the impacts of land use scenarios. Figure 2.24 shows the historical (in 1950) and current land use maps for comparison. Approximately 42% of the Geul subcatchment has experienced changes in land covers since 1950. In 1950, grassland, forest and orchards occupied some 73%, while currently these land cover types occupy approximately 68%. The decrease of 5% is due to the nearly total disappearance of orchards, partly compensated by an increase of grassland acreage and a very small increase in forested area. The arable land remained constant (21%), but the proportion of weeded and mixed crops almost doubled from 29% to 57% (weeded crops from 24% to 43% and mixed crops from 5% to 14%). The urban area almost doubled from 6% in 1950 to 11% in 2000 (Dautrebande et al., 2000).

During the recent three decades, flood mitigating measures have been carried out

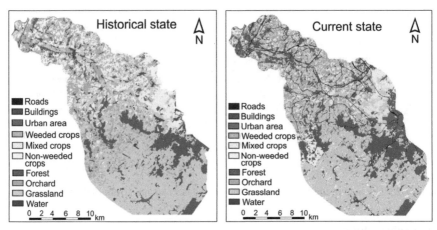

Figure 2.24 Historical (in the 1950s, left) and current (right) land use in the Geul
          subcatchment (source: adapted from Dautrebande *et al.*, 2000). Weeded crops
          include maize, potatoes, beets, vegetables; non-weeded crops mainly include
          wheat; buildings refer to those in agricultural area.

in the Geul subcatchment and the effects of "flowing storage", a natural solution by
slowing down the water and increasing the storage capacity (e.g. restoring
meanders) have become visible in the Geul between Partij and Mechelen
(Arknature, 2000). According to the information collected from Water Quality
Board Limburg (Zuiveringschap Limburg, The Netherland), the Geul subcatchment
has three retention areas with a total capacity of more than 9,000 m^3 (Agor, 2003).

2.7.8 Summary

- The forest area in the Meuse basin upstream of Borgharen has been relatively
  stable in its extent during the 20th century. Nevertheless, the forest type and
  forestry management practices have experienced notable changes. The
  percentage of coniferous forests in the area shows an obvious increase by at
  least 30%, while the percentage of deciduous forests has dropped by the same
  rate.
- Intensification of agricultural land and rapid urbanisation are the most
  important land use changes in the Meuse basin in the second part of last
  century. The extent of agricultural intensification culminated between the
  1950s and the 1970s. The process was linked to mechanisation and
  motorisation of agriculture, and modifications in crop rotation systems.
- Urbanisation has been continuously increasing since the 1950s. Nevertheless,
  the coverage is still limited relative to the entire basin area.
- When catching a general picture of the land use changes in the past century,
  one should recognise the importance of changes in land use management
  (without obvious changes in broad land cover categories). Unfortunately,
  some of these changes are often poorly documented and difficult to illustrate.
- In response to requirements for both navigation and flood protection, many
  hydraulic measures have been taken place in the Meuse river network. Except
  for the Eau d'Heure reservoir complex, most reservoirs in the upper branches
  of the tributaries upstream of Borgharen are relatively small. The total
  drainage area of the reservoirs accounts for only 7% of the basin area.

2.8 Statistical trend detection methods

2.8.1 Introduction

Detection of inconsistency or non-homogeneity in the observation series commonly employs statistical tests, either parametric or non-parametric. The choice between the two families of tests is based on the expected distribution of data involved. If data is normally distributed, parametric tests are usually selected. If data is expected to be non-normally distributed, non-parametric tests are preferred. Transformations are sometimes used to make data more normally distributed, prior to performing a parametric test (Helsel and Hirsch, 1992). McCuen (2003) summarized the advantages and disadvantages of non-parametric tests and parametric tests. Cavadias (1995) and Peterson et al. (1998) have listed representative homogeneity tests that have been employed in the climatological and hydrological data over the recent decades. Some of the homogeneity tests depend on meta-data, while the others are purely statistical. The presence of a single significant test result may only be weak evidence of change. If more tests are significant, then this provides stronger evidence of change, unless they are very similar, in which case multiple significance is not an extra proof of change. However, application of more than one test to the data may make interpretation of results complex (Kundzewicz and Robson, 2004). Given the differences in the underlying assumptions of the tests and the possible influence of change in the catchment condition, it is usually difficult to compare and particularly to combine the results of the different tests (Cavadias, 1995). Use of distribution-free testing methods are recommended for hydrological data which are often strongly non-normal distributed (Kundzewicz and Robson, 2004).

In the literature, three measures of correlation are commonly used to measure monotonic relationships between two continuous variables: Kendall's τ, Spearman's ρ and Pearson's r (Helsel and Hirsch, 1992). The first two are non-parametric and have similar power in detecting a trend (Yue et al., 2002). Application of the two tests for trend analysis in hydrological and climatological series can be found, e.g. Conley and McCuen (1997), Voortman (1998), Douglas et al. (2000), Gonzalez-Hidalgo et al. (2001), Burn and HagElnur (2002), and Beighley and Moglen (2002). The more commonly used Pearson's correlation coefficient r is a measure of linear correlation, one specific type of monotonic correlation. It assumes that the data follow a bivariate normal distribution. This assumption rules out the use of r when the data have increasing variance. Thus, r is often not useful for describing the correlation between untransformed hydrological variables (Helsel and Hirsch, 1992). However, a discussion of the normality distribution of variables is frequently ignored in the applications. Detection of long-term, linear trends is affected by a number of factors, including the size of trend to be detected, the time span of available data, and the magnitude of variability and autocorrelation of the noise in the data. The number of years of data necessary to detect a trend is strongly dependent on, and increases with, the magnitude of variance and autocorrelation coefficient of the noise. For most expected environmental changes, several decades of high-quality data will be needed before such changes will be detectable (Weatherhead et al., 1998). The difficulty with a trend is in deciding whether or not any trend quantified will continue unchanged into the future, since it may only be part of a long cycle of change (Shaw, 1991).

A number of tests are also available for determining shifts in the mean and variance of time series. The parametric t-test and the non-parametric Mann-Whitney test are commonly used in hydrological and climatological time series for determining a shift in the mean, when the point of change is known. When the point of change is not known, change point tests are powerful to detect subtle changes, for example, the Pettitt test (Pettitt, 1979) and the Standard Normal Homogeneity Test (SNHT) (Alexandersson, 1986; Alexandersson and Moberg, 1997). The Pettitt test is a rank-based test for cases in which the initial distribution is unknown. Generally, the rank-based change point tests are simple to use and interpret, and also robust in the presence of outliers. Application of the Pettitt test in climatological studies can be found, e.g. Bárdossy and Caspary (1990), Tomozeiu *et al.* (2002). The SNHT test is a parametric test by assuming normality of the station series, often used to identify sudden shifts (in the mean and variance) and gradual trends in the station series relative to the reference series of surrounding sites (e.g. Jaagus and Ahas, 2000; Wijngaard *et al.*, 2003 and Können *et al.*, 2003). The commonly used cumulative deviation test, also by assuming normally distributed data, is relatively powerful for a change point that occurs towards the centre of the time series (Kundzewicz and Robson, 2004). The detection of linear trends can be confounded when abrupt changes occur in the records. When an abrupt change occurs near either the beginning or the end of the record, the trend tests may often suggest insignificant changes in the data series (McCuen, 2003). Weatherhead *et al.* (1998) reported that abrupt changes in the data can strongly impact the number of years necessary to detect a given trend, increasing the number of years by as much as 50%; the impact of the shift on trend detection strongly depends on its relative location in time in the data set.

Identification of trends and change points in data sets may be complicated by autocorrelation (also referred to as persistence or serial correlation). A pre-whitening approach (see Appendix A) is often used to remove the serial correlation, prior to applying a statistical test (e.g. Burn and Hag Elnur, 2002). Using the Monte Carlo simulation, Yue and Wang (2002) demonstrated that removal of serial correlation by pre-whitening can effectively remove the serial correlation and eliminate the influence of the serial correlation on the Mann-Whitney test (for detecting a shift in median). Yue and Pilon (2003) proposed a trend-free pre-whitening procedure to eliminate the effect of serial correlation on the Mann-Kendall test. However, if the serial correlation coefficient for lag 1 (r_1) is too large or the time series too short, pre-whitening the data affects any trend that may be present in the data (Douglas *et al.*, 2000).

2.8.2 Procedure and tests applied

This study emphasises identification of non-homogeneities (both gradual trends and abrupt changes/shifts) in the hydrological, hydro-meteorological and climatological time series. A rigorous procedure was developed and the statistical tests used for detecting trends and change points were determined:

  i) First, the Spearman rank correlation method was applied to evaluate the absence of linear trend, at a significance level of 5%.
  ii) Then, change point analysis was performed using both the non-parametric Pettitt test and the parametric SNHT test without a reference series. A critical probability level of 80% was chosen for acceptance of significant change points in the Pettitt test results, while a critical significance level of 90% tabulated in Alexandersson and Moberg (1997) was used in the SNHT test results. Meta-data, if available, were used to justify the change point results. For application of the SNHT test, the data series were also tested for

normality and, if necessary, transformed to fit a two-parameter Lognormal distribution (LN) or a three-parameter Lognormal distribution (LN3). Certainly, the requirement of normal distribution was not always satisfied. In this case, only the Pettitt test was applied.

iii) Once the candidate change points were identified, split-record tests on means (two sided *t*-test) and variances (*F*-test) were carried out to quantify the magnitude of the changes and to test the significance (at a significance level of 5%) of the differences in the subsets before and after the change point.

iv) Serial correlation of the data series was examined. When the r_1 (lag-1 serial correlation coefficient) value was significant (at a significance level of 5%), the pre-whitened series was evaluated.

The principles of some statistical methods applied are briefly described in Appendix A. The statistical analyses in this study were facilitated by the DATSCR program (Dahmen and Hall, 1990) for the Spearman's rank test and the SPELL-Stat program (Guzman and Chu, 2003) for serial correlation analysis, the Pettitt test and the SNHT test. It should be mentioned that in the Pettitt test results the first year before a change is recognised as a change point, while in the SNHT test results the first year after a change is taken as a change point. To express consistency, the change point of a time series mentioned in the text refers to the first year after a change.

The change point test results in this study show that less significant change points were detected by the SNHT test. The main reason is possibly application of the relatively high critical level in the test. In addition, hydrological data often have characteristics that make the application of parametric tests either inappropriate or inefficient. Nevertheless, in most cases, the significant change points indicated by the SNHT test for the selected variables are the same as or similar to the results given by the Pettitt test. Therefore, evaluation of the long-term changes in the observation series in the latter chapters was mainly based on the change point results identified by the Pettitt test.

In the interpretations, the change point years identified should not be considered in an absolute way. As mentioned previously, the effects of climate variability and land use change usually lead to gradual changes in streamflows. One should realise that the change point year identified in a record period marks the start of a change, but does not necessarily marks an abrupt change in river flows. In addition, when environmental changes did occur, due to highly variable nature of the data series and different lengths of records analysed, the exact time of the identified change for different variables and different records may slightly vary.

3   Changes in the discharge regimes

3.1 Introduction

During the period of observations, a wide variety of influences may cause the discharge time series either inconsistent or non-homogeneous. Inconsistency is characterised by a change in the amount of systematic error associated with the recording of the data, while non-homogeneity results from natural or man-made changes to the environment (Shaw, 1991). This chapter aims to examine the temporal changes in the discharge regimes of the Meuse river and the selected tributaries based on their available longest records. Both mean flows (section 3.2) and extremes including floods (sections 3.3 to 3.4) and low flows (section 3.5) in the rivers are examined, with more attention paid to the floods. In addition, maximum moving-average streamflows and timing of occurrence of flood peaks are also included (section 3.3). The identification of abrupt changes or shifts (regarding the mean values) in the observation series is emphasised in the statistical analyses, since it is very useful when searching for possible reasons behind the changing hydrological behaviour of the river system. Comparatively, the commonly performed detection of linear trends receives less attention.

3.2 Change in the mean flows

3.2.1 Analysed variables and test periods

Mean flows in the Meuse river and the selected tributaries are investigated on different temporal scales, including:
- annual average discharge (denoted as AAD), defined as the arithmetic average of the daily discharges over a hydrological year spanning from November to October;
- average discharge for each season (denoted as WIND for the winter period from December to February, SPRD for the spring period from March to May, SUMD for the summer period from June to August and AUTD for the autumn period from September to November), defined as the arithmetic mean of the monthly average discharges in that season;
- average discharge for each month (denoted as JAND for January, FEBD for February, MARD for March, APRD for April, MAYD for May, JUND for June, JULD for July, AUGD for August, SEPD for September, OCTD for October, NOVD for November and DECD for December), defined as the arithmetic mean of the daily discharges in that month.

In the Meuse basin, the hydrological year is usually defined from October to September, based on the precipitation excess (precipitation-evaporation). In this study, a slightly different definition of the hydrological year is used, based on the flow regime of the Meuse river during the year (see Figure 3.1). In Figure 3.1, a similar flow regime is also found in the selected tributaries, with the difference that the Rur, the Jeker and the Geul show less fluctuation. The relatively even pattern of the runoff distribution in the Jeker and the Geul during the year is attributable to the effect of groundwater storages provided naturally by extensive chalk or limestone aquifers in the catchment areas. The runoff distribution in the Rur has been affected by the regulation of the reservoirs in the subcatchment, as demonstrated in the latter sections. In this study, the hydrological year is designated by the ending year, e.g. the first hydrological year for Monsin is recorded as 1912, which starts from 1 November, 1911 and ends on 31 October, 1912. Hence, the length of a time series (in the hydrological year) is reduced from n to n-1 years.

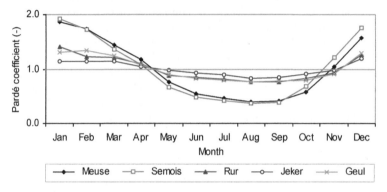

Figure 3.1   Flow regimes of the Meuse (near Monsin, 1911–2000) and the selected tributaries Semois (at Membre, 1929–2002), Rur (at Stah, 1953–2001), Jeker (at Nekum, 1965–2001) and Geul (at Meerssen, 1953–2001), described in terms of the dimensionless Pardé coefficients (defined as ratios of the average monthly discharges of 12 months to the average annual discharge, respectively).

Analyses of mean flows in the Meuse river are mainly based on the reconstructed Monsin record. The gauged Borgharen record is only used in the analysis of annual average discharge (AAD) for comparison with AAD near Monsin. It should be mentioned that the analysed discharge records for the Meuse and the selected tributaries have varying lengths. The time series of mean flow variables for the Meuse are examined within two time frames: the long test period starting from 1911/1912 and the short test period starting from 1950. For the tributaries excluding the Jeker, the entire periods of record are tested: 1929/1930–2002 for the Semois, 1953/1954–2001 for the Rur and the Geul. Visual inspection of the derived discharge time series for the Jeker suggests an inconsistency around 1965, possibly arising from an inappropriate correction of the low flow hard copy data. Therefore, the shorter period 1965/1966–2001 is analysed for the Jeker. The outcomes of statistical trend analyses can be fairly sensitive to the selected length of record. Therefore, the statistical results for the Meuse obtained from the short test period are used for comparison with the statistical results for the tributaries, in case the different statistical results are obtained for the Meuse from the two test periods. Tu *et al.* (2004a) has reported the main results of mean flows in the Meuse river.

3.2.2 Annual mean flows

The correlograms of AAD in the Meuse and the tributaries all show significant r_1 values, with an extreme case of AAD in the Jeker which shows rather strong serial correlation for short lags (e.g. $r_1 = 0.77$ and $r_2 = 0.47$, far beyond the upper critical level of 0.33).

Analysis of AAD in the Meuse at Borgharen indicates a significant decrease since 1932 (with a probability of 0.86 for the year 1931 in the Pettitt test result) for the long test period (1912–2002), which became statistically insignificant after pre-whitening. No significant change point was detected for AAD near Monsin (see Figure 3.2), regardless of the record length. Therefore, the decrease in AAD at Borgharen around 1932 (about 44 m³/s on average) could be largely explained by canal water extractions between Liège and Borgharen (see Figure 2.4). Consistent findings were obtained in the selected tributaries, with no significant change in their AAD series (see Figure 3.2). Variation of AAD in the Jeker shows a sign of time lag. The results of the Spearman test for trend suggest no significant trend for AAD in the Meuse (near Monsin) and the tributaries over their defined test periods. From the above statistical test results, it appears that AAD in the Meuse (near Monsin) and the selected tributaries have been relatively stable.

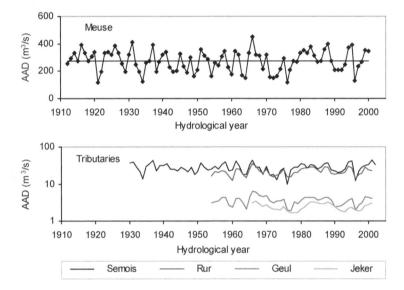

Figure 3.2 Time series of AAD in the Meuse (near Monsin, 1912–2000) and the selected tributaries Semois (at Membre, 1930–2002), Rur (at Stah, 1954–2001), Geul (at Meerssen, 1954–2001) and Jeker (at Nekum, 1966–2001). The black line (for the Meuse) represents the mean value over the entire test period. The logarithmic scale is used for the tributaries.

3.2.3 Seasonal mean flows

All seasonal flow series of the Jeker show strong serial correlation for lag 1 or in general for short lags. Other serially-correlated seasonal flow series include WIND in the Rur, and SPRD and SUMD in the Geul.

Analysis of the Monsin record on a seasonal basis indicates a relative increase in SPRD from 1978 to 1989 and a significant decrease in AUTD from 1933 onwards for the Meuse (see Figure 3.3). No statistically significant change was detected for WIND and SUMD in the Meuse, regardless of the record length. The change point results obtained from the selected tributaries partly support the findings from the Meuse, depending on the site and the season. The most consistency appears in AUTD in the tributaries, i.e. no significant change over their defined test periods, which appears to support the stable nature of AUTD in the Meuse since 1933.

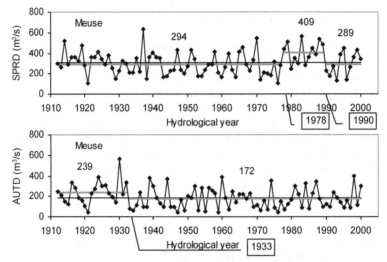

Figure 3.3   Change point results of SPRD and AUTD in the Meuse river (near Monsin, 1912–
2000). The black thin lines represent the mean values over the entire study
periods. The grey bold lines represent the mean values over the corresponding
data subsets. The numbers indicate the mean values for the corresponding data
subsets. The same definitions apply to the subsequent figures.

A large similarity was also observed for the temporal variabilities of SPRD in the
tributaries except for the Geul, with a slight difference that the secondary change
point year 1990 for the Meuse became statistically insignificant in the Pettitt test
results for the Semois and the Rur. Furthermore, the *t*-test results for the two rivers
indicate insignificance of the relative increases in SPRD after 1977/1978.
Comparatively, the discharge in the Semois appears more variable in most seasons.
A statistically significant increase since 1980 (with a probability of 0.87 for the year
1979 in the Pettitt test result) was detected for WIND in the Semois (see Figure 3.4),
but with no evident change for WIND in the other tributaries. SUMD in the Semois
was also found to show a distinct decrease around 1967 (see Figure 3.4), while
SUMD in the Geul shows a statistically significant decrease around 1989.

None of the seasonal flow series of the Meuse and the tributaries exhibits a
statistically significant trend, except for SUMD in the Semois showing a significant
downward trend which is related to the distinct decrease noted above.

3.2.4 Monthly mean flows

Almost all monthly flow series of the Jeker show serial correlation. Other serially-
correlated monthly flow series include JAND in the Rur, and MARD, MAYD,
JUND, JULD and DECD in the Geul.

In the Meuse river, pronounced changes were found for MARD, SEPD and
NOVD. The increase identified for MARD in the river occurred in 1978, same as the
first change point year identified for SPRD in the river. SEPD in the river shows the
same change point year of 1933 as that for AUTD in the river, while NOVD in the
river gives a varying change point year of 1945. Figure 3.5 illustrates the change
point results of MARD, SEPD and NOVD in the Meuse. For the short test period
from 1950 to 2000, DECD in the river shows a change point year of 1978 with a

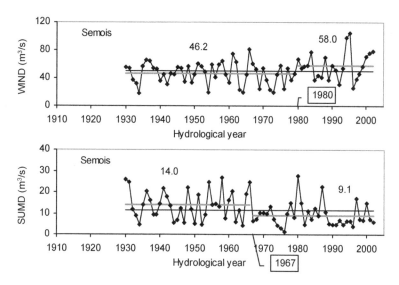

Figure 3.4 Change point results of WIND and SUMD in the Semois (at Membre, 1930–2002).

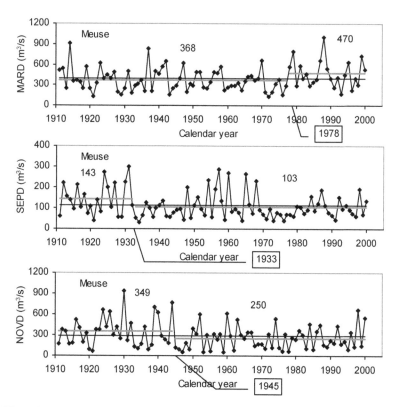

Figure 3.5 Change point results of MARD, SEPD and NOVD in the Meuse river (near Monsin, 1911–2000).

probability of 0.79 in the Pettitt test result, not statistically significant with the critical probability level of 0.80.

Figure 3.6 summarizes the discernible changes in the selected tributaries using bar charts. Comparatively, the Semois shows more non-homogeneous monthly series. The notable increase of WIND in the Semois found in the last couple of decades seems to primarily result from the change in JAND. The increase shown in MARD in the Semois from 1978 onward appears to support the finding from MARD in the Meuse. The decrease identified for SUMD in the Semois around 1967 was evident in JUND and AUGD with slightly varying change point years (1974/1973). The monthly flow series of the tributaries that are not included in Figure 3.6 do not show statistically significant changes and therefore are considered relatively stable (homogeneous) over the analysed periods.

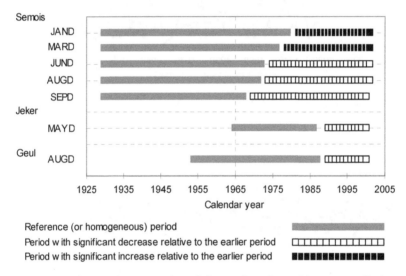

Figure 3.6   Bar chart of segmentation of time series of monthly average discharges in the selected tributaries Semois (at Membre, 1929–2002), Rur (at Stah, 1953–2001), Jeker (at Nekum, 1965–2001) and Geul (at Meerssen, 1953–2001), based on the dominating change point years.

The Spearman test results indicate significant upward or downward trends only for JUND, JULYD, AUGD and SEPD in the Semois, APRD in the Rur and AUGD in the Geul. None of the monthly flow series of the Meuse shows a significant trend.

3.3 Change in the high flows

3.3.1 Analysed variables and test periods

Generally two types of floods can be distinguished in the Meuse basin: winter floods and summer floods. Winter floods are often associated with abundant precipitation and wet soils in the entire basin, while summer floods are generally influenced by thunderstorms and often relatively small in size due to a large soil moisture deficit and localised precipitation. Extreme floods in the Meuse often occur in the winter season. Figure 3.7 depicts the monthly distribution of annual daily maxima in the

Meuse (at Borgharen) as well as in the selected tributaries. In this figure, use is made of relative frequencies so that the distributions are independent of the absolute values of the monthly frequencies and the record lengths. It should be mentioned that the lag time of peaks (after intensive precipitation events) in such small tributaries as the Jeker and the Geul is not more than one day (e.g. only four to ten hours for the Geul) and, thus, the daily maxima (based on the daily mean discharge records) being analysed in this study probably can not reflect the dynamic runoff characteristics in the two rivers. Unfortunately, the instantaneous peak series of good quality for both rivers were not available for this study.

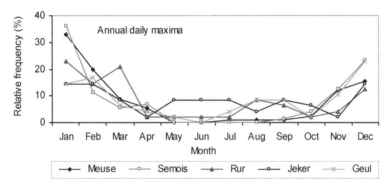

Figure 3.7 Monthly distribution of annual daily maxima in the Meuse river (at Borgharen, 1912–2002) as well as in the selected tributaries Semois (at Membre, 1930–2002), Rur (at Stah, 1954–2001), Jeker (at Nekum, 1954–2001) and Geul (at Meerssen, 1954–2001).

In this study, daily maxima and maximum average discharge over a certain period for the entire hydrological year (November–October), the winter (November–April) and the summer (May–October) were analysed separately. Annual maximum consecutive k-day moving average discharge (denoted as AMAXkD, where $k = 1, 3, 5, 7, 10, 15$ and 30 days) was derived for each river, as well as maximum consecutive k-day moving average discharge for both the winter (denoted as WMAXkD, where $k = 1, 3, 5, 7, 10, 15$ and 30 days) and the summer (denoted as SMAXkD, where $k = 1, 3, 5, 7, 10$ and 15 days). For the Meuse, the analyses concentrate on the Monsin record, while for Borgharen only AMAXD, WMAXD and SMAXD are examined. It should be realised that the maximum k-day time series for each station are not independent and exhibit more or less correlation. In addition, the date of occurrence of WMAXD for each river is investigated, by analysing the number of days counted from the beginning of each hydrological year, i.e. 1 November. The periods for analysing flood flows in the Meuse and the tributaries (excluding the Jeker) remain the same as those for AAD in the rivers (see section 3.2.2). Analyses of the flood flows in the Jeker are based on the entire record from 1954. Tu *et al.* (2004b and 2005a) has shown the main results of high flows in the Meuse river.

The relationship between flood magnitude and frequency is usually described by a flood frequency curve based on extreme value theory. It basically requires that the discharge data series analysed should be homogeneous. In section 3.3.5, WMAXD in the Meuse (at Borgharen) for different sub-periods are analysed based on the Gumbel distribution (see Appendix B), intending to know whether the relationship of concern has significantly altered during the 20th century. Fitting the observation data to alternative distributions, e.g. the Lognormal and exponential distributions, has been attempted and nothing additional can be learned from them. The frequency models used for computing the design discharge of the Meuse (at Borgharen) are presented in Chbab (1995) and Parmet *et al.* (2001).

3.3.2 Daily maxima

Among the daily maximum series of the Meuse and the tributaries, AMAXD and WMAXD in the Semois, AMAXD in the Rur and SMAXD in the Jeker show significant serial correlation for lag 1. The r_1 values for AMAXD and WMAXD in the Meuse over the past 91 years are very close to the upper 95% confidence limits.

The change point results for Borgharen suggest significant increases in both AMAXD and WMAXD around 1984. Their probabilities in the Pettitt test results are high (above 0.90). Further segmentation of the time series reveals another significant change around 1971. No significant change was found in SMAXD for Borgharen. The test results for Monsin are almost consistent with those for Borgharen, except that AMAXD near Monsin shows the dominating change point year around 1980 for the long test period. Figure 3.8 illustrates the temporal variabilities of AMAXD in the Meuse together with those of AMAXD in the selected tributaries.

For the Semois and the Rur, pronounced increases in AMAXD and WMAXD were found since 1980 or 1979. The daily maximum series of the Semois also shows instability in variances. Further segmentation of the time series of AMAXD and WMAXD indicates a significant change around 1971 for the Rur. In the Jeker, SMAXD shows a statistically significant increase around 1977, while the increase in AMAXD since 1981 is not statistically significant (indicated by the *t*-test result). As an exceptional case, the Geul presents evident downward shifts in both AMAXD and WMAXD around 1971, accompanied by reduced temporal variations. Compared with the "dry" decade of the 1970s, AMAXD and WMAXD in the Geul during the last two decades still appear a little higher. The temporal variabilities of AMAXD in the tributaries are shown in Figure 3.8.

The Spearman test results indicate significant (upward) trends only for AMAXD and WMAXD in the Semois and SMAXD in the Jeker, even after pre-whitening the series. The upward trends of AMAXD and WMAXD in the Meuse, SMAXD in the Semois and AMAXD in the Jeker are associated with high probabilities (>90%). The downward trends of AMAXD and WMAXD in the Geul are not statistically significant.

3.3.3 k-day maximum runoff

Investigation of the *k*-day maximum runoff ($k \geq 3$) in the Meuse (near Monsin) shows significant increases after 1984, but only for AMAX3D and WMAX3D over the long test period, and for AMAX5D and WMAX5D over the short test period. For increasing *k* (e.g. $k \geq 5$), the effects of AMAXD and WMAXD in the river appear to have been smoothed out due to averaging. Considering the large basin scale and the extensive channel routing, the volume of storm runoff in the Meuse could be regarded as little changed.

In the tributaries, the notable changing patterns identified for the daily maximum series or similar changing patterns are also seen in most of the associated *k*-day maximum series with $k \geq 3$. There seems little change in the volume of summer storm runoff in the Meuse as well as in the tributaries except for the Jeker.

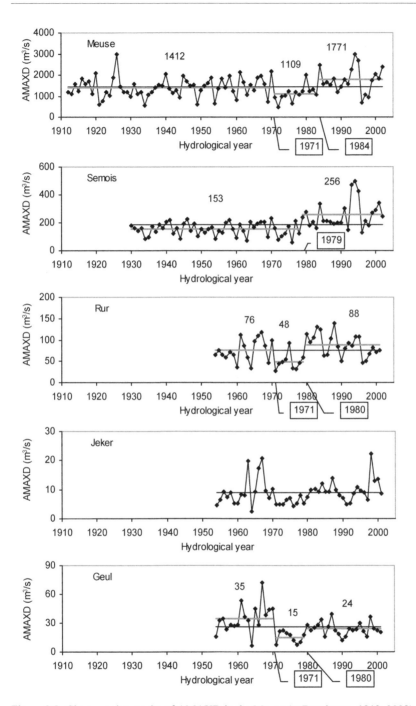

Figure 3.8 Change point results of AMAXD in the Meuse (at Borgharen, 1912–2002) and the selected tributaries Semois (at Membre, 1930–2002), Rur (at Stah, 1954–2001), AMAXD in the Jeker (at Nekum, 1954–2001) and Geul (at Meerssen, 1954–2001), based on the multiple change point years.

3.3.4 Date of occurrence of winter daily maxima

Among the time series of days for occurrence of WMAXD (denoted as day-for-WMAXD) in the Meuse and the tributaries, only the one for the Rur exhibits serial correlation for lag 1. The day-for-WMAXD series of the Meuse shows two significant change point years with very close probabilities in the Pettitt test result, 1933 (0.919) and 1941 (0.922). There seems to be a transition period between these two years. After 1942, the date of occurrence of WMAXD in the Meuse appears to have significantly postponed for more than 20 days on average (see Figure 3.9). No obvious change was found for the day-for-WMAXD series of the tributaries except for the Jeker. It seems that the occurrence of WMAXD in the Jeker has notably advanced for about one month since 1990.

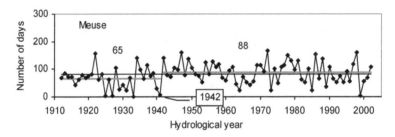

Figure 3.9 Change point result of day-for-WMAXD for the Meuse (at Borgharen, 1912–2002), counted from the beginning of each hydrological year, i.e. 1 November.

3.3.5 Frequency analysis of winter daily maxima

In section 3.3.2, it was found that WMAXD in the Meuse (at Borgharen) has significantly increased since 1984. Therefore, the Gumbel plots of WMAXD for Borgharen before and after 1984 are compared. Equal lengths of 19 years are taken to facilitate the comparison. Because the 1970s have a relatively "dry" climate, a "normal" period 1952–1970 is chosen. Furthermore, the early "undisturbed" (i.e. reference) period 1912–1930 is added. Figure 3.10 shows their Gumbel plots.

In Figure 3.10, the points of the data subset from the "undisturbed" sub-period (1912–1930) are located within the 95% confidence limits and nearly follow the straight fitted line. This is an indication that the Gumbel distribution is a reasonable fit to the data set of WMAXD for Borgharen. The impact of the largest flood in 1926 (3000 m^3/s) on the fitting was checked through comparison of the Gumbel plots (1912–1930) with and without this flood event (not shown). This appears to have little influence on the frequency analysis results of floods smaller than 2000 m^3/s. Relative to the early sub-period, the data subsets from the latter two sub-periods (1952–1970 and 1984–2002) clearly show some different statistical properties. The points in the middle parts of their Gumbel plots are partly located outside of the 95% confidence limits of the Gumbel plot for 1912–1930. Moreover, the points in the middle part of the Gumbel plot for 1984–2000 also lie outside of the 95% confidence limits of the Gumbel plot for 1952–1970 (not shown). The apparent bias (i.e. upward departure) of floods in the middle part of the Gumbel plot for 1984–2002 suggests that the floods with the magnitude in the range of roughly 1300 to 2500 m^3/s occurred more frequently (statistically significant) in the last two decades. For the larger floods for instance the last two flood events in the upper tail

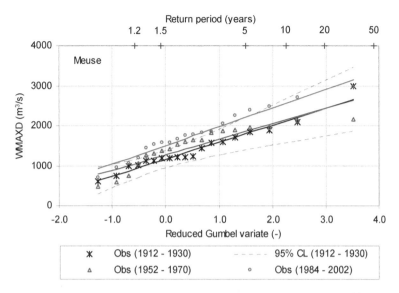

Figure 3.10 Gumbel plots of WMAXD in the Meuse (at Borgharen) for different sub-periods: 1912–1930, 1952–1970 and 1984–2002. The straight lines are the reference lines for the three sub-periods, respectively. The 95% confidence limits (CL) for 1912–1930 are added.

(of the Gumbel plot for 1984–2002), one can not say that the frequency of occurrence of floods has statistically significantly increased because the flood peaks are still within the wide range of the 95% confidence limits for 1912–1930.

Another concern is the bias of small floods in the lower tails of the Gumbel plots for 1952–1970 and 1984–2002 (Figure 3.10), which exhibits a downward departure from their straight fitted lines. Whether the bias of small floods influences the Gumbel fitting of medium and large floods was checked. Figure 3.11 shows the Gumbel plots of WMAXD for Borgharen after omitting small floods ($< 1000 \ \text{m}^3/\text{s}$) in the lower tails. In the figure, the original data subsets are also plotted for illustration. It clearly shows that the two data subsets with removing small floods conform the Gumbel distribution well. Their points in the middle and upper parts of the Gumble plots are also located closely (just a slight decrease in return period) to the points of the Gumbol plots without removing small floods. This indicates that the bias of small floods in the Gumbel plots in Figure 3.10 would not substantially affect the frequency analysis results of larger floods.

3.4 Change in the peaks-over-threshold

3.4.1 Definition of peaks-over-threshold

Annual maximum (AM) series and peaks-over-threshold (POT) series (also known as partial duration series) are two types of flood data series. Both types of data series make assumptions of statistical independence. There are generally more values for analysis of the POT series (depending on the threshold) than of the AM series, but there is more chance of the peaks being related and the assumption of independence

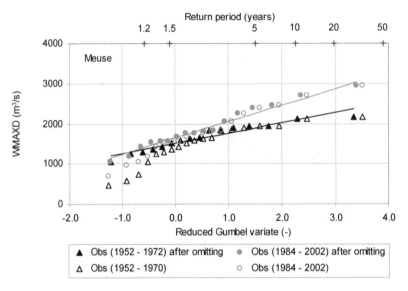

Figure 3.11 Gumbel plots of WMAXD in the Meuse (at Borgharen) with removing several points of small values in Figure 3.10 (three points for 1952–1970 and two points for 1984–2002). The original data sets are also plotted for illustration.

is invalid. For sufficiently long records, it may be prudent to consider all the major peaks and then the threshold is chosen so that the number of peaks equal to and exceeding this level corresponds with the number of years of record, but not necessarily one peak in each year. The data series derived in this way is often called the annual exceedence (AE) series, a special case of the POT series. The AM series and the POT series form different probability distributions. For return periods exceeding 10 years, the differences between the two types of data series are negligible and the AM series is the one usually analysed (Shaw, 1991). In this sub-section, the long-term change in the POT series of the Meuse river is investigated after analysing the AM series (i.e. AMAXD) in section 3.3. The main results have been previously reported by Tu *et al.* (2006). Increases in peak discharges from any headwater catchment can have little effect on downstream peaks because of the routing and desynchronisation that normally occur (Brooks *et al.*, 1997). Therefore, the tributary Semois in the Ardennes, for which a long discharge record was available, is investigated as a parallel case. As stated in section 1.2.2, land use changes have a more pronounced effect on lower peaks than on extreme discharges. Analysis of the POT series is therefore particularly useful to provide an indication of how land use changes may have affected more frequent floods, e.g. floods with a return period of smaller than five years (by defining the threshold).

The criterion for independence is a minimum time between flood peaks, which is dependent on the catchment size. The winter (November–April) peak discharge (WMAXD) in the Meuse (based on the Monsin record) was found to best correlate with the antecedent precipitation depth (based on the MeuseRLP record) for seven to ten days (e.g. $r = 0.77$ for the case of seven days) on wet soils (see section 6.3.1). Therefore, a minimum period of ten days between adjacent peaks was considered appropriate to ensure independence of the peaks in the Meuse. In most cases, the period between two floods was much larger than ten days.

Another crucial decision is the choice of a threshold for the POT analysis. In the probability distribution function applied to the Meuse for Borgharen, the flood frequencies higher than once every 25 years were estimated using the general Pareto distribution applied to the POT series with a threshold level of 1300 m^3/s, whereas the lower frequencies were estimated using the Gumbel, Pearson III, Lognormal and the Exponential distributions applied to the AM series (see Parmet *et al.*, 2001). Based on the performance functions for Borgharen given in Parmet *et al.* (2001), the 1.5-year flood, the 5-year flood and the 10-year flood are 1473 m^3/s, 1898 m^3/s and 2143 m^3/s, respectively. The term "flood" in the Meuse for Borgharen was previously used if the discharge exceeded 1450 m^3/s (Berger, 1992). In this study, two major factors were considered when defining a threshold level for the POT events in the Meuse: *i*) Frequently-occurring small peaks in the river should be included, so as to benefit for evaluating the effects of land use changes in the basin. *ii*) Sufficient peaks should be obtained, so as to achieve statistical significance.

Three different threshold levels were finally applied for the Meuse: 800 m^3/s, 1217 m^3/s and 1500 m^3/s. The obtained peak series were denoted as POT_{800}, POT_{1217} and POT_{1500}, respectively, in which the subscripts written next to POT represent the different threshold levels. The threshold levels of 800 m^3/s and 1500 m^3/s are just above the 90th percentile (labelled as Q90 and calculated as 775 m^3/s) and the 99th percentile (labelled as Q99 and calculated as 1482 m^3/s) of winter half-year daily discharges at Borgharen, respectively. The threshold level of 1217 m^3/s is the lowest peak value (on 29 January, 1941) of the largest 91 peaks in the 91 years of record, hence resulting in the AE series for the Meuse. Starting from the highest peak of 3000 m^3/s, the POTs were selected from the discharge hydrographs in the two sub-records (1911–1925 vs. 1926–2002) segmented on the date of occurrence of the highest peak (on 1 January, 1926), with a minimum time interval of ten days between the adjacent peaks. Figure 3.12 illustrates the procedure of selecting the POTs. Based on the 91-year record for Borgharen, totally 219 POT_{800} peaks, 91 POT_{1217} peaks and 50 POT_{1500} peaks were obtained.

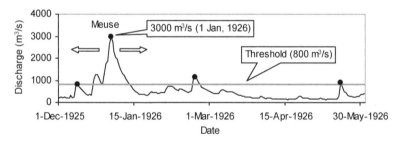

Figure 3.12 Illustration of the procedure for selection of POTs in the Meuse (at Borgharen). Black dots represent the selected POTs equal to or exceeding 800 m^3/s.

Similarly, a minimum time period of seven days between adjacent peaks and three different threshold levels (90 m^3/s, 158 m^3/s and 180 m^3/s) were used to define the POTs in the Semois. The associated peaks were denoted as POT_{90}, POT_{158} and POT_{180}, respectively. The threshold levels of 90 m^3/s and 180 m^3/s correspond to respectively the Q90 and the Q99 of winter half-year daily discharges at Membre. The threshold level of 158 m^3/s (on 1 November, 1932) was the lowest peak value of the largest 73 peaks in the 73-year record. In total, 230 POT_{90} peaks, 73 POT_{158} peaks and 52 POT_{180} peaks were obtained for the Semois.

3.4.2 Analysed variables and test periods

The derived POT series for the Meuse and the Semois form the basis for further statistical trend analysis. Table 3.1 lists the basic characteristics of the POT series. In the table, the basic characteristics of the AM series are added for comparison. For the Meuse, the AM series and the POT_{1217} series share 56 common peaks, accounting for about 62% of their total 91 peaks. For the Semois, the AM series and the POT_{158} series share 45 common peaks, similarly accounting for 62% of their total 73 peaks. Furthermore, Figure 3.13 depicts the monthly distribution of the POT events together with the AM events for both rivers, in which the relative frequencies were used to remove the effect of the different sample sizes. Clearly, the majority of the POT events occurred in the winter half-year. Winter POT events (denoted as WPOT) are therefore of special interest, concerning their generally great magnitudes and meaningful sample sizes, while summer POT events in both rivers receive less attention in this study.

Table 3.1 Basic statistics of the POT series for the Meuse river (at Borgharen) and the Semois river (at Membre), compared with the basic statistics of the AM series.

Statistics	Meuse (1912–2002)				Semois (1930–2002)			
	AM	POT_{800}	POT_{1217}	POT_{1500}	AM	POT_{90}	POT_{158}	POT_{180}
Number of peaks	91	219	91	50	73	230	73	52
Number of peaks per year	1	2.4	1	0.6	1	3.2	1	0.7
Mean (m^3/s)	1445	1254	1632	1866	188	149	216	236
Standard deviation (m^3/s)	513	409	371	350	83	61	66	69

In the change point analyses, the following POT-related variables and aspects are considered: *i*) frequency of the POT events (i.e. number of peaks) in the hydrological year; *ii*) magnitude of the POT events (i.e. peak discharge) in time; and *iii*) seasonality of the POT events. Various POT series are examined over their entire record lengths. Further segmentation is also appropriately considered to provide insights into possible short-term changes. Nevertheless, caution should be exercised in the interpretation, since they may possibly be associated with random property of the data and thus sometimes have little practical implication.

3.4.3 Frequency

The annual frequency of the POT_{800} events in the Meuse shows an obvious increase after 1979. A similar pattern is also evident for the POT_{90} events in the Semois (see Figure 3.14). For the larger POT events in both rivers, the POT series contain many zero-values and therefore the Pettitt test was not conducted. Their temporal changes were inspected visually. No obvious increase in the last couple of decades was found for the annual frequencies of the POT_{1217} events and the POT_{1500} events in the Meuse. Nevertheless, compared with the earlier periods of the same length, the latter period still contains more flood events. The annual frequencies of the POT_{158} eventsand the POT_{180} events in the Semois appear to have notably increased since 1979. Figure 3.15 shows the percentages of flood events in the rivers for the different periods. The period after 1979 appears to be "flood-richer" for both rivers.

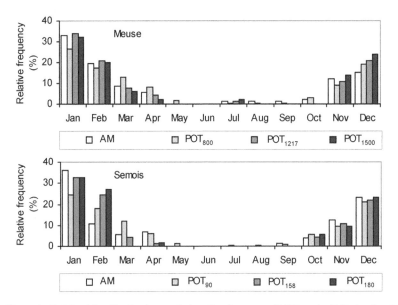

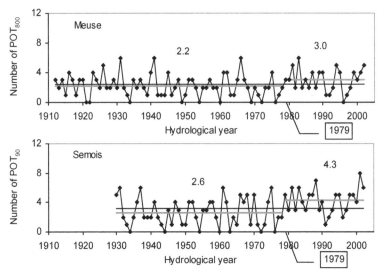

Figure 3.13 Monthly distributions of the flood events (POT and AM) in the Meuse (at Borgharen, 1912–2002) and the Semois (at Membre, 1930–2002), respectively. The subscripts written next to POT represent the defined thresholds (in m^3/s). The same definitions apply to the subsequent figures.

Figure 3.14 Change point results of the annual frequencies of POT_{800} events in the Meuse (at Borgharen, 1912–2002) and POT_{90} events in the Semois (at Membre, 1930–2002).

For the POT events with peak discharges between the lower and higher thresholds, i.e. the $POT_{800-1500}$ events in the Meuse and the POT_{90-180} events in the Semois, the Pettitt test did not detect any significant change point in their annual frequency series. Their inter-annual variations appear to be too large.

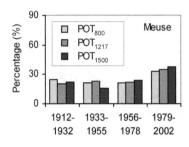 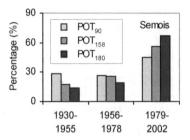

Figure 3.15 Percentages of the different POT events in the Meuse and the Semois for the different sub-periods.

3.4.4 Magnitude

Over the long test period from 1912 to 2002, the magnitude of the POT_{800} events in the Meuse shows a distinct decades-long fluctuation, characterised by higher values roughly from the mid-1940s to the end of the 1960s and from 1984 until present, and lower values during the early 30 years from 1912 to 1944 and the 1970s. The magnitude of the POT_{1217} events appears to have significantly increased since 1984. The magnitude of the POT_{1500} events shows a change point located on 23 December, 1991, with a probability of 0.87 in the Pettitt test result. Nevertheless, the *t*-test result indicates insignificance of the increase after that change point (i.e. since 1993). Figures 3.16 shows the change point results of the magnitudes of the different POT events for the Meuse.

For the magnitudes of all defined POT events in the Semois, the *F*-test values on the variances of the subsets before and after the change points are statistically significant, which is mainly caused by a few extreme floods occurring in the winter of 1993/1994 and 1995. Nevertheless, the overall increase in the last couple of decades is still evident. Figures 3.17 shows the change point results for the Semois.

Analyses of the WPOT events for both rivers were also carried out. The magnitude of the $WPOT_{800}$ events in the Meuse shows a similar fluctuation pattern as exhibited in the magnitude of the POT_{800} events. The magnitude of the $WPOT_{90}$ events in the Semois shows an increase most likely around 1977. Since the larger WPOT events in both rivers are almost the same as their corresponding POT events, there is little difference between their change point results.

No obvious change was found for the magnitude of the $POT_{800-1500}$ events in the Meuse (see Figures 3.16) as well as for the magnitudethe POT_{90-180} events in the Semois (see Figures 3.17). Likewise for the $WPOT_{800-1500}$ events in the Meuse and the $WPOT_{90-180}$ events in the Semois.

3.4.5 Flood regime

Changes in the flood regimes of the Meuse and the Semois will be illustrated on a seasonal basis, to retain a meaningful sample size for the larger flood events. Comparison of the seasonal distributions of the POT events in the Meuse before and after 1979 suggests an insignificant increase in the relative frequency in spring and an insignificant decrease in autumn after 1979. The changes were most notable in the months of March, April and November, during which most of small POT events (e.g. < 1200 m^3/s) occurred. Analyses of the $POT_{800-1500}$ events in spring and autumn

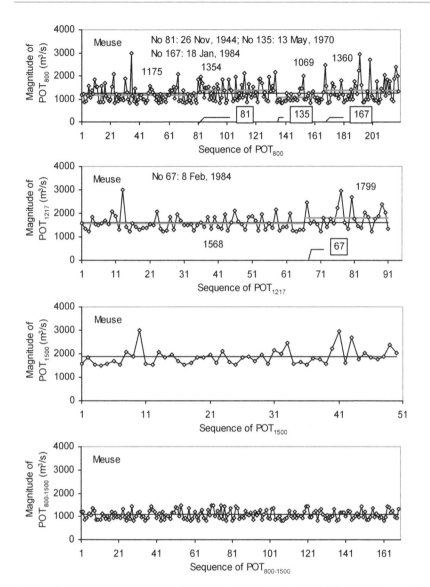

Figure 3.16 Change point results of the magnitudes of different POT events in the Meuse (at Borgharen, 1912–2002).

for the Meuse indicate no significant change in their magnitudes.

The marked changes in the flood regime of the Semois are the notable increase in the relative frequencies of larger floods in winter and the decrease in autumn after 1979. For the POT_{90-180} events in the river, the relative frequency in spring also shows an increase after 1979, meanwhile compensated by a decrease in winter. Nevertheless, no significant change point was detected in the magnitude of the POT_{90-180} events in the river for each season. Figure 3.18 depicts the seasonal distributions of the $POT_{800-1500}$ events in the Meuse and the POT_{90-180} events in the Semois before and after 1979, respectively.

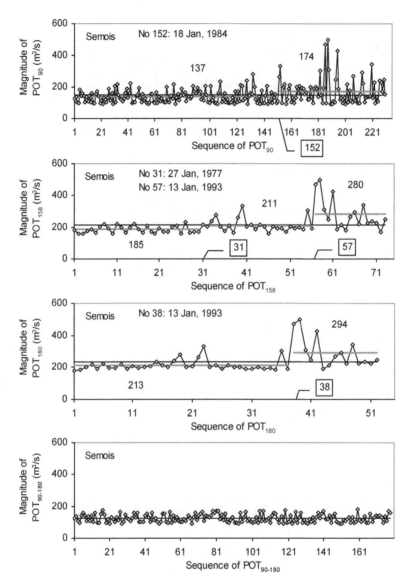

Figure 3.17 Change point results of the magnitudes of different POT events in the Semois (at Membre, 1930–2002). Note that the variances of the POT_{158} subset and the POT_{180} subset after 1993 are quite large.

3.5 Change in the low flows

3.5.1 Analysed variables and test periods

Low flows in the dry season are generally composed of baseflow from groundwater reservoirs, which are significantly affected by catchment geology and long-term climatic fluctuations (Sokolovskii, 1971; Smakhtin, 2001). In the upper section of

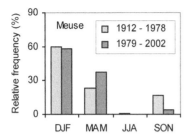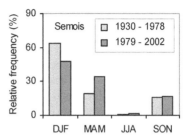

Figure 3.18 Comparison of the seasonal distributions of $POT_{800-1500}$ events in the Meuse and POT_{90-180} events in the Semois for the different sub-periods. Four seasons: winter = DJF, spring = MAM, summer = JJA, and autumn = SON.

the Meuse basin, the lithology is largely formed by limestone and thus in dry periods the discharge of the Lorraine Meuse is relatively high. In the middle section of the Meuse basin, a great part consists of poorly permeable rocks and thus has limited groundwater storage capacity. After a period of little precipitation in the summer and autumn, the discharge of the river near Monsin may drop below 60 m³/s, which can cause the problem of water shortage with respect to socio-economic, ecological and environmental aspects in the downstream part. More downstream the discharge is sustained by the Rur, some small brooks and groundwater (Berger, 1992). Low flow conditions in the Meuse are affected by water extractions (especially between Liège and Borgharen, see Figure 2.4), reservoirs (especially in the Rur) and the operation of weirs for navigation. In the Jeker and the Geul, groundwater is a major contributory factor to low flows during the dry periods. The low flow regime of a river can be analysed in a variety of ways. Different types of low flow indices can be derived either from the flow duration curve, considering low flow spells, or from the frequency analysis of the annual minimum series of low flows (Shaw, 1991). For example, the 95% exceedence flow (i.e. the percentile Q95) in the flow duration curve during the considered period is usually suggested as a low flow measure, and the change in the Q95 may be analysed annually or seasonally to illustrate variations of the low flow. Low flows are normally experienced during a drought (also referred to as a low flow spell in the literature) and feature one element of the drought, i.e. the drought magnitude. Other characteristics of hydrological droughts, which have been frequently studied, include duration, deficit volume (severity) and time of occurrence (e.g. Hisdal and Tallaksen, 2000; Hisdal *et al.*, 2001).

This study takes a close look at the magnitude of summer low flows in the Meuse as well as in the selected tributaries, defined by summer half-year (May–October) minimum consecutive 10-day moving average discharge (denoted as SMIN10D). There are some practical reasons for giving preference to analysing minimum average flows for a few days or longer period. For example, averaging of daily observations can eliminate the day-to-day variations due to water management, and an analysis based on such time series is less sensitive to measurement errors. However, in the majority of cases, there is no big difference between daily and several-day low flows (Smakhtin, 2001). For an analysis of low flows, it is preferable to have natural discharge records unaffected by major extractions. The Monsin record corrected for extractions is therefore used for the Meuse. Figure 3.19 shows the time series of SMIN10D for Monsin, in comparison with SMIN10D for Borgharen. The discrepancy between the two SMIN10D series is caused by the average estimates of water extractions between Liège and Borgharen. There could be an overestimation of SMIN10D near Monsin for the dry summers of the 1970s due to application of the fixed extraction estimates in the reconstructed record. During the 89 years of record, the monthly distribution of the occurrence of SMIN10D near Monsin was found as follows: once in May, twice in June, 13 times in July, 18 times in August, 30 times in September and 25 times in October. Almost all summer low flows in the Meuse occurred in the period from July to October, with preference to late summer and early autumn. The inaccuracy resulting from the corrections made for the water extractions becomes less with the larger period of averaging, e.g. on a seasonal or annual basis. Therefore, the average discharge for the period from July to October (denoted as Jul-OctD) near Monsin is also used to define summer low flows in the Meuse. Figure 3.19 also compares the time series of Jul-OctD near Monsin with Borgharen. The global patterns of inter-annual variations shown by SMIN10D and Jul-OctD near Monsin appear to a large extent similar to Borgharen. One point should be mentioned that the devision of a hydrological year used in this study partitions the yearly low flow period of the Meuse, causing the "low flows" in November not to be included in the analyses. The monthly distribution of the occurrence of annual minimum 10-day moving average discharge (denoted as AMIN10D) near Monsin was similar to the monthly distribution of SMIN10D and was found as follows: ten times in July, 13 times in August, 28 times in September 17 times in October, 19 times in November and twice in December. The high chance of occurrence of AMIN10D in the transition month of November appears to be associated with a large soil moisture deficit after dry extended summer months. However, there is little difference between SMIN10D and AMIN10D in the Meuse (near Monsin) with respect to the magnitude, since the low flows are largely sustained by the relatively stable baseflows. The test periods for summer low flows in the Meuse and the tributaries are the same as those for AAD in the rivers (see section 3.2.2). Tu *et al.* (2005c) has specifically presented the main results of low flows in the Meuse river.

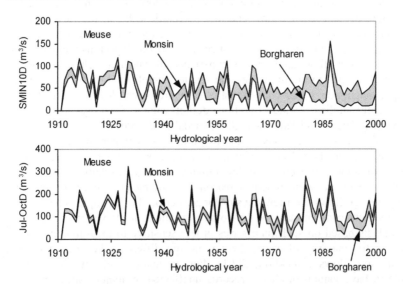

Figure 3.19 Time series of SMIN10D and Jul-OctD near Monsin compared with Borgharen (1912–2000), respectively. The grey belt represents the average estimates of water extractions from the Meuse between Liège and Borgharen.

Analysis of low flow spells and deficit volumes is not carried out in this study. These two low-flow indices are defined based on certain thresholds. For a constant threshold of 60 m³/s near Monsin, during the considered 89 years there are more than 2700 low flow days with a daily discharge less than 60 m³/s, in which about 2400 days (accounting for 88%) occurred in the summer half-year. Figure 3.20 shows summer low flow days in the Meuse near Monsin.

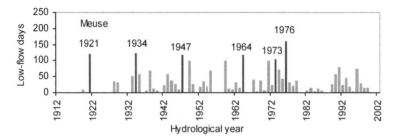

Figure 3.20 Summer low flow days of the Meuse with daily discharges near Monsin less than 60 m³/s. In the particularly dry years 1921, 1934, 1947, 1964, 1973 and 1976, represented by the black bars, the low flow situation lasted more than 100 days.

3.5.2 Magnitude

Table 3.2 summarizes the serial correlation coefficients of lag 1 to lag 3 (i.e. r_1, r_2 and r_3) for the SMIN10D series. There is indication of significant inter-annual persistence for all SMIN10D series which results from slowly reacting aquifer systems and/or long-term climatic fluctuations. For the SMIN10D series of the Geul and the Jeker, pre-whitening (using Eq. 1-14 in Appendix A) is inefficient to remove the lag-1 serial correlation.

Table 3.2 Summary of the serial correlation coefficients of lag 1 to lag 3 (r_1, r_2 and r_3) for the SMIN10D series.

River	Meuse	Semois	Rur	Geul	Jeker
Station	Monsin	Membre	Stah	Meerssen	Nekum
Period	1912–2000	1930–2002	1954–2001	1954–2001	1966–2001
r_1	**0.213**	**0.229**	**0.284**	**0.670**	**0.799**
r_2	0.114	0.215	0.244	**0.367**	**0.536**
r_3	0.034	0.152	**0.452**	0.154	0.313
95% CL	±0.208	±0.229	±0.283	±0.283	±0.327

Note: Bold values are significant at a 95% confidence level (CL). The sign "±" indicates the upper and lower bounds of CL.

Analysis of SMIN10D in the Meuse near Monsin indicates a downward shift around 1933 for the long test period, without obvious changes either for the sub-period 1933–2000 or for the short period 1950–2000. The significance of the change point year still remains even after pre-whitening. From 1933 onwards, SMIN10D in the Meuse appears to have significantly decreased by about 23 m³/s on average (see Figure 3.21), which agrees well with the result for AMIN10D near Monsin. Similar variation was also discernable for Jul-OctD near Monsin, with a significant decrease of 32 m³/s after 1933 (see Figure 3.21).

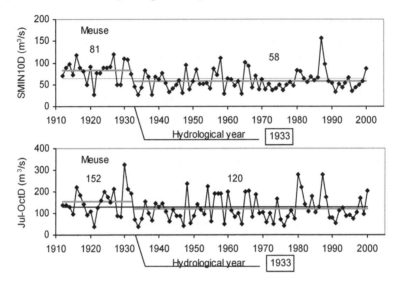

Figure 3.21 Change point results of SMIN10D and Jul-OctD in the Meuse (near Monsin, 1912–2000).

Since the rates of water extractions from the Meuse are quite large in comparison with the low flows occurring near Monsin, suspicion of the data quality arose: Does the accuracy of the reconstructed Monsin record justify statistical analysis of low flows in the Meuse upstream of Liège? In other words, does the decrease identified in SMIN10D near Monsin result from underestimation of the water extraction from the Meuse? Therefore, a comparison of the Monsin record has been made with the relatively long record (1960–1997) for the station Chooz (see Figure 2.1) on the Ardennes Meuse. The Chooz record has been measured by MET-SETHY and was available from the FRIEND data base. The drainage area upstream of Monsin is about twice that upstream of Chooz (about 10,120 km²). Controlled by comparable hydrogeological characteristics, the year-to-year fluctuation patterns of SMIN10D near Monsin and at Chooz appear similar and their magnitudes defined in depth are also close except for a slight discrepancy over a few years just before and after 1980 (see Figure 3.22). With the increase of averaging period such as from July to October, the discrepancy in magnitude (Jul-OctD) between the two stations becomes negligible (not shown). The overall consistency between the two SMIN10D series as well as between the two Jul-OctD series infers that the reconstructed Monsin record could provide reliable information on low flows in the Meuse. Unfortunately, the comparison made above does not include the early sub-period prior to 1933.

Figure 3.23 illustrates the change point results of SMIN10D in the selected tributaries. SMIN10D in the Semois shows a distinct decrease (by 50%) around 1971, accompanied by reduced year-to-year variations. SMIN10D in the Rur shows a significant increase around 1965, followed by a decrease around 1988. The change points detected for SMIN10D in the Geul (around 1989) and the Jeker (around 1992) lie close to the end of their test periods. The short-term decreases shown in the last decade appear to be part of a strong cycle in their SMIN10D series.

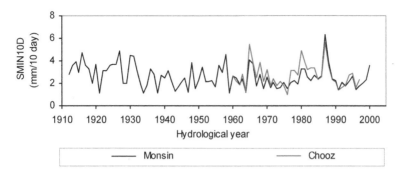

Figure 3.22   Comparison of the Monsin record with the Chooz record regarding SMIN10D.

The notable drops in SMIN10D in the Meuse and the Semois cause significant downward trends over their whole study periods. No clear trend is evident in SMIN10D in the Rur, the Jeker and the Geul. The downward trend in Jul-OctD in the Meuse is not statistically significant.

3.6 Discussion

In sections 3.2 to 3.5, a range of hydrological variables reflecting different flow states (i.e. mean flows, high flows and low flows) of the Meuse river and the selected tributaries have been examined for detection of their abrupt changes and linear trends. The results of the Spearman test often suggest absence of significant linear trends due to relatively high variability of the hydrological data series, whereas the results of the Pettitt test in many cases indicate the existance of significant abrupt changes. In the following paragraphs, the trend results with emphasis on the abrupt changes/shifts are discussed with reference to the quality of the discharge data used. Other influential factors along with the relevant literature will be elaborated in Chapter 6 where the rainfall-runoff relationships of the study areas are addressed.

In spite of the relative stability on an annual scale, various average discharges of the Meuse river and the selected tributaries show either significant or insignificant changes on a seasonal scale. There seems a common evidence of a short-term "jump" in SPRD in the rivers (excluding the Geul) from the late-mid 1970s to the end of the 1980s. The notable changes in other seasons are site specific, including AUTD in the Meuse (decrease around 1933), WIND (increase around 1980) and SUMD (decrease around 1967) in the Semois, and SUMD in the Geul (decrease around 1989). The change point results from the monthly series of the rivers provide some additional insight to the observed seasonal changes, although these results are in general considered less reliable due to the relatively larger variations in the time series. For example, the change shown in SPRD in the Meuse appears to be largely related to the change in MARD. A similar relation is seen in the Semois. The high flows and the low flows of the Meuse river and the selected tributaries were also found to show striking changes during the 20th century. The trend results of two types of flood magnitude series derived from the daily discharge record (1911–2002) for Borgharen are largely consistent, both (annual/winter maximum series vs. POT magnitude series with two different thresholds of 800 m^3/s and 1217 m^3/s)

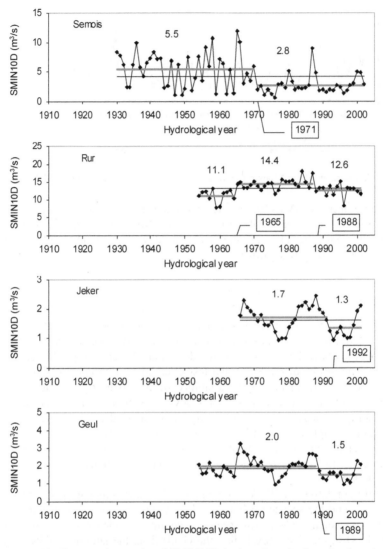

Figure 3.23 Change point results of SMIN10D in the selected tributaries Semois, Rur, Jeker and Geul. The test periods are the same as those for AAD in the rivers.

demonstrating that the magnitude of winter (November–April) floods in the Meuse has significantly increased since 1984. Simultaneously, the trend results of the POT frequency series suggest an increased frequency of winter floods (e.g. > 800 m³/s) since 1979, with more frequent occurrence of small floods (e.g. < 1200 m³/s) in spring. The trend result of the POT magnitude series for the Meuse with the higher threshold of 1500 m³/s does not show a significant increase, probably related to its sampling strategy. The flood trend of the Meuse river is supported by the frequency analysis result of winter half-year daily maxima (WMAXD) for Borgharen based on the Gumbel distribution. The annual/winter maximum flood series for the selected tributaries (excluding the Jeker and the Geul) exhibit concurrent increases (statistically significant) in the last couple of decades. The changes identified in the summer half-year low flows (SMIN10D) of the rivers show less consistency. It is worth noting that the results of the Spearman test indicate insignificance for most of the linear trends in the flow series.

It must be recognised that there is a considerable uncertainty about the findings obtained for the river basins under study. The changes for the derived hydrological variables could be related to inconsistency of the discharge record from that site. Changes in gauge location, instrument and observation frequency, and the stage-discharge relations can produce abrupt changes in flow rates during the transition period and some of these changes may have influenced specific discharge characteristics in the river. For example, compared with extreme high or low flows, the abrupt changes identified in the monthly to annual mean flows derived from the same daily discharge records are in general less sensitive to inconsistency or errors. In this study, the available meta-data on the discharge records are therefore used to discuss the impact of inconsistency on the change point results. As introduced in section 2.2.1, the corrections to the discharge data observed at Borgharen for water extractions from the Meuse river resulted in the reconstructed Monsin record. Because the amounts of water extractions (ranging from 20 m^3/s to 50 m^3/s on average) are not accurately known particularly during the year, the derived hydrological variables from the same Monsin record have different data qualities, depending on how large the extraction rates are in comparison with the analysed discharges occurring near Monsin. Generally, there is a difficulty in making accurate low flow measurements and the low flow series of rivers are sensitive to data errors. It appears that the distinct downward shifts around 1933 for the related variables of AUTD (by 67 m^3/s, see Figure 3.3), Jul-OctD (by 32 m^3/s, see Figure 3.21) and SMIN10D (by 23 m^3/s, see Figure 3.21) near Monsin could be somewhat related to the inaccurate corrections made for the water extractions, similar to the case of AAD at Borgharen (see section 3.2.2). However, it should be realised that the inaccuracies of the extraction rates are very unlikely in the order of 50 m^3/s. Moreover, from comparison of the Jul-OctD series for Monsin and the upstream station of Chooz (see Figure 3.22), it appears that the Monsin record for the latter sub-period after 1933 is reliable. Therefore, the decreases in AUTD, Jul-OctD and SMIN10D in the Meuse (near Monsin) might be partly or largely caused by other factors. Meanwhile, there is a suspicion of the data quality of the Monsin record for the earlier sub-period prior to 1933. The high flow series for Borgharen and Monsin are much less sensitive to the inaccuracies of the extraction rates because the amount of water extracted is relatively small compared to flood discharges, but subject to uncertainty in the measurements, particularly in the cases of extreme floods. For all dominant changes identified for the Meuse, the impact of the changes in the instrument and the measurement frequency around 1975 for Borgharen (see section 2.2.1) appears to be negligible. The marked decreases for the related variables of SUMD (around 1967) and SMIN10D (around 1971) in the Semois occurred near the time of the changes in the instrument and the observation frequency (around 1968) for Membre (see section 2.2.1) and also coincide with the start-up year (1967) of the small Vierre reservoir in the tributary Vierre (see section 2.7.7). The Vierre reservoir seems too small to cause a significant decrease of the low flows. Therefore, it is more reasonable that SUMD and SMIN10D in the Semois are perhaps subject to the influence of measurement system change of discharge (at Membre), which, however, can not directly explain the increase of WIND in the river aound 1980. The discontinuity (around 1960) of the daily discharge record for Stah (see section 2.3.4) seems to have little impact on the findings obtained for the Rur. The data quality of the daily discharge record for the Jeker is subject to the influence of combining two different data sources (with the discontinuity around 1972) and a correction made for the sub-period from 1991 due to underestimation (see section 2.3.3). The decrease of SMIN10D in the Jeker around 1992 seems to be somewhat associated with the underestimation of the record, but is more likely due to some regional factors because the adjacent tributaries such as the Rur and the Geul also show similar decreases. The lag in the temporal variations of SMIN10D in the rivers results from slowly reacting aquifer systems and/or long-term climatic fluctuations. The dominant decrease for AMAXD and WMAXD in the Geul is striking, in contrast with the apparent increases in the Meuse and the other selected tributaries in the last couple of decades. The change point year 1971 for the Geul coincides with the change in the recording system (and also the station move) for Meerssen (see section 2.3.2). There seems to be a link between these two changes, which needs to be confirmed through further investigation. Moreover, during the last decades flood mitigating measures (e.g. restoring meanders and increasing retention capacity; see De Laat and Agor, 2003) were carried out in the Geul subcatchment.

3.7 Concluding remarks

Based on the statistical results and the discussion above, the following concluding remarks can be drawn:
- The annual average discharges (AAD) in the Meuse river (near Monsin) and the selected tributaries have been relatively stable during the analysed periods. However, there exist some evidences of seasonal changes in the rivers, e.g. a short-term "jump" in the spring average discharge (SPRD, mainly March) from the late-mid 1970s to the end of the 1980s. The other seasonal changes are site specific.
- The flood peak discharges (i.e. AMAXD and WMAXD) in the Meuse river (at Borgharen) and the selected tributaries (excluding the Jeker and the Geul) have significantly increased in the last two decades of the 20th century (most likely since 1980). The timing of occurrence of winter maximum floods in the Meuse river has been significantly postponed since the early 1940s. There also exists evidence of more small floods (e.g. $< 1200 \ m^3/s$) in the Meuse river in spring.
- The summer low flows (i.e. SUM10D) of the Meuse river (near Monsin) and the selected tributaries show notable abrupt changes with varying change point years.

To give a proper interpretation of the changes identified for the derived hydrological variables, further investigation is certainly required. Climate variability and land use change are the common examples of changes that can cause non-homogeneity in the discharge records. The main findings regarding these issues for the Meuse river and the selected tributaries constitute the subsequent chapters (Chapters 4 to 7). The final conclusions are drawn in Chapter 8.

4   Changes in the precipitation regimes

4.1 Introduction

In the literature, statistical trend studies of observed regional (or local) precipitation changes are numerous. For the Meuse basin and its vicinity, several researchers have recently examined global trends in the precipitation observations (using either station series or basin-average series) for different variables, e.g. WL (1994), KNMI (1999), Gellens (2000), Pfister *et al.* (2000), De Wit *et al.* (2001) and Vaes *et al.* (2002). De Wit *et al.* (2001) observed a small increase in the annual (November–October) and winter (November–April) precipitation depths from a basin-average precipitation record (1911–1998). Gellens (2000) found significant trends in extreme winter (October–March) *k*-day precipitation depths but none in extreme summer (April–September) *k*-day precipitation depths for the climatological network (165 stations) of Belgium (1951-1995). However, analysis of a few stations with long records starting from 1910 showed no significant trend. Vaes *et al.* (2002) examined the trends in the precipitation (1898–1997) at Uccle (Belgium) based on peaks-over-threshold (POT) values. For small aggregation levels (i.e. less than one day) there was a small decrease in extreme rainfall events over the century, while for large aggregation levels there was a more explicit increase in extreme rainfall events. However, the author could not conclude that the trends are significant. So far, a general picture of the main features of precipitation changes in the entire basin is still lacking. In this chapter, the available long records of daily areal precipitation for the Meuse basin and two selected (Rur and Geul) subcatchments are systematically analysed, intending to interpret the temporal changes identified for the streamflow variables in Chapter 3. In addition to the analysis of changes in monthly, seasonal and annual precipitation totals (section 4.2), this study also includes an analysis of changes in the annual and seasonal precipitation extremes (section 4.3) and the frequency of precipitation events (section 4.4). The statistical trend results are discussed (section 4.5), followed by some concluding remarks (section 4.6).

4.2 Change in the precipitation totals

4.2.1 Analysed variables and test periods

The total precipitation amounts in the Meuse basin and the Rur and Geul subcatchments are investigated on different time scales ranging from month to year, including:
- annual precipitation total (denoted as APT) in the hydrological year;
- precipitation total for each season (denoted as WINP for winter, SPRP for spring, SUMP for summer and AUTP for autumn), computed by summing the monthly precipitation totals in that season, together with the percentage of seasonal precipitation total in the annual sum (of four seasonal precipitation totals) for each season, which may filter out the possible effect of instrumental changes at all rain gauging stations on the observation values;

- precipitation total for each month (denoted as JANP for January, FEBP for February, MARP for March, APRP for April, MAYP for May, JUNP for June, JULP for July, AUGP for August, SEPP for September, OCTP for October, NOVP for November and DECP for December), computed by summing all daily precipitation values in that month.

Figure 4.1 depicts the precipitation regimes in the study areas. Application of three different precipitation records (see section 2.4.2) for the Meuse basin results in similar fluctuation patterns, except for the MeuseSP record showing a small difference in May and June. Compared with the selected two subcatchments, the Meuse basin appears to have a wetter mid-autumn to winter due to influence of the topography factor and a drier late-spring to summer. In this study, the precipitation analyses for the Meuse basin are mainly based on the MeuseLP record. The MeuseSP record and the MeuseRLP record, which became available much later, are examined only with respect to APT for comparison with the MeuseLP record. For a similar reason, the Geul5P record (see section 2.4.3) is used for the Geul subcatchment but without a comparison with the Geul7P record. In Agor (2003), a great consistency was found between these two records with respect to a similarly defined APT. The test periods of the derived precipitation variables are consistent with those of the mean flow variables defined for the Meuse (1911/1912-2002 for the long test period and 1950-2002 for the short test period), and the tributaries Rur and Geul (1953/1954–2001). Tu *et al.* (2004a) has summarized the main results for the Meuse basin.

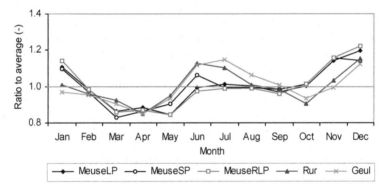

Figure 4.1 Distribution of monthly precipitation values in the Meuse basin (based on the MeuseLP, 1911–2002; the MeuseSP, 1928–1998; the MeuseRLP, 1911–2000) as well as in the Rur (1953–2001) and Geul (based on the Geul5P, 1953–2001) subcatchments, represented as proportions of the average of the 12 monthly values in that catchment area.

4.2.2 Annual precipitation totals

The r_1 values of the APT series for the Meuse basin and the two selected subcatchments are all within the confidence limits set at 95% confidence level, indicating the absence of significant serial correlation for lag 1.

The Pettitt test result of APT based on the MeuseLP record shows a change point year around 1979, with probabilities of 0.84 for the long test period and 0.86 for the short test period. The statistical significance of the candidate year was also confirmed by the split-record tests on means and variances. However, different conclusions were obtained from the MeuseSP (1930–1998) and MeuseRLP (1912–

2000) records, both of which suggest insignificance of the year 1979 in the Pettitt test results. To clarify the possible influence of including the most recent wet years (i.e. 2000 and 2001) on detectability of the change point in APT over the basin, APT based on the MeuseLP record was tested again for periods of different length 1930–1998 and 1912–2000 as well as for different short periods 1950–1998 and 1950–2000. None of the test periods suggests significance of the year 1979 in the Pettitt test result. Therefore, the conclusions seem sensitive to lengthening of the test periods with the most recent wet years. Figure 4.2 presents the Pettitt test results of APT based on the MeuseLP record for illustration. In this study, the conclusion obtained from this record over the long test period 1912–2002 is preferred. Relative to the earlier sub-period 1912–1979, APT in the Meuse basin appears to have just significantly increased (by 78 mm, nearly 8%) after 1980 (see Figure 4.3).

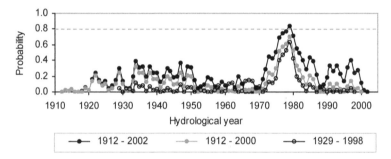

Figure 4.2 Pettitt test results of APT based on the MeuseLP record for three periods (1912–2002, 1912–2000 and 1930–1998). The 80% probability level is indicated in the graph.

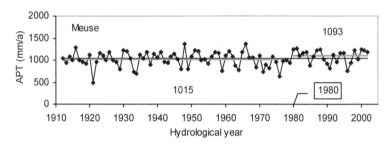

Figure 4.3 Change point result of APT in the Meuse basin (based on the MeuseLP, 1912–2002).

For APT in the Rur and Geul subcatchments, the Pettitt test detected the same change point year of 1979, with probabilities of 0.88 and 0.52 (below the critical level of 80%), respectively. The further analysis by the t-test indicates insignificance (but close to the critical level) of the candidate year for APT in the Rur subcatchment. It appears that the increase identified in APT in the Meuse basin around 1980 is less evident or even indiscernible in the two subcatchments. Nevertheless, a similar percentage of increase in APT from 1980 onwards (nearly 9%) was still obtained for the Rur subcatchment.

No significant linear trend in APT in the Meuse basin as well as in the two subcatchments could be detected, regardless the length of the test period.

4.2.3 Seasonal precipitation totals

Among all seasonal precipitation series, only WINP in the Meuse basin for the short test period and WINP in the subcatchments show serial correlation. Their r_1 values are just statistically significant.

Analyses of the MeuseLP record on a seasonal basis indicate some changes for SPRP and SUMP. Over the considered 91 years, SPRP in the Meuse basin shows a similar short-term "jump" as identified for SPRD in the Meuse river (Figure 3.3), during which SPRP has increased by about 30% relative to the other two sub-periods (Figure 4.4). SUMP in the Meuse basin shows a decrease around 1967, but significant only for the short test period (Figure 4.4). WINP and AUTP in the Meuse basin appear to have been relatively stable over the past century. The two variables show change point years around 1979 or 1980 (over the long test period) with low probabilities (0.47 and 0.50, respectively) in the Pettitt test results. Additional analyses of the percentages of seasonal precipitations give slightly varying change points for the percentages of SPRP and SUMP in the basin, with no obvious changes occurring in the percentages of WINP and AUTP. The jump in the percentage of SPRP covers the period 1964–1989. The percentage of SUMP over the long test period shows a significant decrease since 1973, suggesting a tendency towards drier summers.

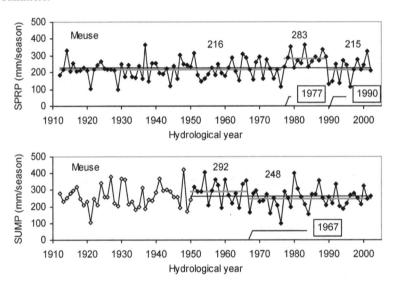

Figure 4.4   Change point results of SPRP (1912–2002) and SUMP (1950–2002) in the Meuse
basin.

In the Rur subcatchment, significant changes were found for WINP and SPRP. WINP in the subcatchment shows a significant increase by 21% after 1980 (see Figure 4.5). After pre-whitening, the change point year (1979) in the Pettitt test result became just statistically insignificant (with a probability of 0.79). SPRP in the subcatchment shows a similar temporal pattern as observed in the Meuse basin (Figure 4.5). The Pettitt test also detected a significant change point around 1971 (with a probability of 0.86) for SUMP in the subcatchment. Nevertheless, the *t*-test result indicates insignificance of the change point. The percentage of decrease in SUMP in the Rur subcatchment since 1972 was found to be almost the same as the percentage for SUMP in the Meuse basin over the short test period, being about 15%. In the Geul subcatchment, temporal changes in the seasonal precipitations are in general not large enough to make them statistically significant over the analysed period. For example, SUMP in the Geul subcatchment shows the change point year of 1969 with a probability of 0.76 (just below 0.8) in the Pettitt test result.

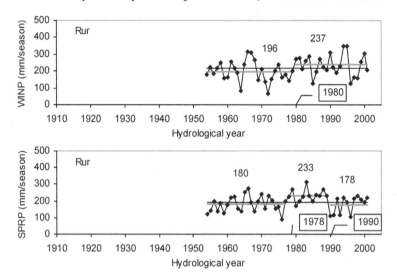

Figure 4.5 Change point results of WINP and SPRP in the Rur subcatchment (1954–2002).

All seasonal precipitation series are subject to upward trends over their test periods, with the exception of the SUMP series showing downward trends. However, none of the linear trends is statistically significant. The long-term upward trend in SPRP in the Meuse basin, mainly caused by the increased precipitation during the 1980s, is associated with a very high probability of 97.4% (just below the upper significant level of 97.5%).

4.2.4 Monthly precipitation totals

None of the monthly precipitation series is serially correlated for lag 1 except for MAYP in the Rur subcatchment ($r_1 = 0.30$, just beyond the upper critical level of 0.28).

Over much of the 20th century, noticeable changes for the monthly precipitation totals in the Meuse basin mainly occurred in March, May and August (see Figure 4.6). In the figure, it is clearly demonstrated that MARP in the basin has experienced a very similar short-term "jump" as observed in SPRP. MAYP in the basin shows a relatively wet sub-period ranging from the 1940s to the 1980s. AUGP in the basin appears to have significantly decreased since 1970, somewhat similar to the decrease found in SUMP for the short test period. In addition, OCTP in the basin shows a significant increase after 1980 for the short test period.

In the Rur and Geul subcatchments, the temporal changes identified for MARP, MAYP and AUGP over the short period of nearly 50 years are consistent with those found in the Meuse basin except for the secondary change point year 1990 being less evident in MARP in the subcatchments. In addition, NOVP in the two subcatchments also show significant changes over the considered period. Figure 4.7 summarizes the discernible monthly changes using a bar graph.

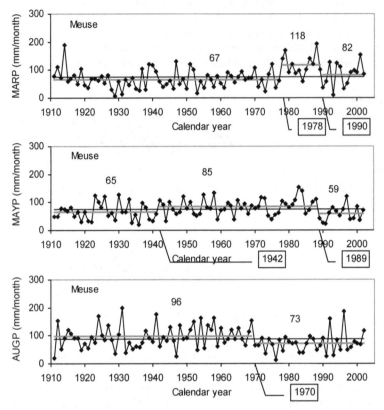

Figure 4.6 Change point results of MARP, MAYP and AUGP in the Meuse basin (1911–
2002).

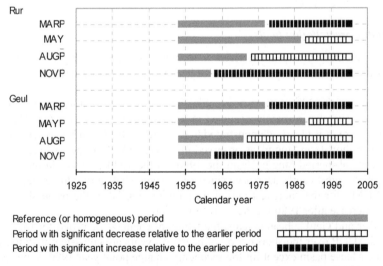

Figure 4.7 Bar graph of segmentation of time series of monthly precipitation totals in the Rur
and Geul subcatchments (1953–2001), based on the dominating change point
years.

Out of all monthly precipitation series for the Meuse basin and the two subcatchments, only the MARP series show significant upward trends, while the AUGP series (excluding AUGP in the Meuse basin for the short test period) show significant downward trends. The linear trends of other monthly series are not statistically significant.

4.3 Change in the precipitation extremes

4.3.1 Analysed variables and test periods

In this sub-section, precipitation extremes in the Meuse basin as well as in the Rur and Geul subcatchments are investigated. The analysed variables include not only annual (November–October) maximum consecutive k-day precipitation depth (denoted as AMAXkP, where k = 1, 3, 5, 7, 10, 15 and 30), but also winter (November–April) maximum consecutive k-day precipitation depth (denoted as WMAXkP, where k = 1, 3, 5, 7, 10, 15 and 30 days) and summer (May–October) maximum consecutive k-day precipitation depth (denoted as SMAXkP, where k = 1, 3, 5, 7, 10 and 15 days). Separation of the time series in winter and summer periods roughly coincides with the different characteristics of two major rain-producing mechanisms in the area (i.e. predominantly frontal in winter and often convective in summer).

For the Meuse basin, the extreme precipitation series derived from the MeuseLP record are once more examined within two time frames: 1912–2002 and 1950–2002. For the Rur and Geul subcatchments, the test periods of their extreme precipitation series are the same as those of their annual precipitation series (1954–2001). It should be realised that the extreme k-day series for various k values at the same site are inter-correlated.

Precipitation extremes are in general highly variable. Table 4.1 provides the recorded maxima of WMAXkP and SMAXkP for various k values over the entire analysed periods. In the table, the winter maxima in the Meuse basin appear to be higher than the summer maxima of the same duration, while the opposite situation is seen in the Rur and Geul subcatchments, which is in line with their regime results as shown in Figure 4.1. Because the summer precipitation distribution in the Meuse basin is generally non-uniform in both space and time, the observed summer extremes are more sensitive to local precipitation variability and thus have weaker spatial correlation. In this context, the temporary variability of summer extremes in the smaller subcatchments is not necessarily in good agreement with that observed in the Meuse basin. Moreover, throughout the year the summer extremes derived for the Meuse basin can also be as high as the winter extremes of the same duration, despite the effect of averaging seven station records.

4.3.2 Annual precipitation extremes

Of all annual maximum k-day precipitation series, only the AMAX10P series for the Meuse basin shows serial correlation for lag 1 over both long and short test periods.

Most of the annual maximum k-day series (excluding AMAXP and AMAX30P) for the Meuse basin show a significant increase around 1980 either over the long test period or over the short test period. For AMAXP in the basin, the Pettitt test detected different change point years for the long test period (1979) and the short

Table 4.1 Maxima of winter and summer maximum consecutive *k*-day precipitation depths (WMAX*k*P and SMAX*k*P, in mm) in the Meuse basin and the selected two subcatchments. The years for the recorded maxima are given in brackets.

Record	Variable	*k*-day						
		1	3	5	7	10	15	30
Meuse	WMAX*k*P	60.7	93.6	110.2	131.8	167.3	208.4	312.1
(1912–2002)		(1941)	(1941)	(1926)	(1995)	(1995)	(1994)	(1994)
	SMAX*k*P	57.2	93.4	110.1	120.4	141.7	168.4	–
		(1966)	(1996)	(1929)	(1929)	(1998)	(1948)	
Rur	WMAX*k*P	53.0	76.3	92.9	103.4	120.2	140.8	230.8
(1954–2001)		(1984)	(1984)	(1992)	(1984)	(1995)	(1994)	(1994)
	SMAX*k*P	57.1	77.9	98.0	105.2	130.9	176.0	–
		(1982)	(1982)	(1984)	(1984)	(1984)	(1969)	
Geul	WMAX*k*P	43.7	65.6	85.4	100.1	106.4[1]	143.0	217.1
(1954–2001)		(1984)	(1984)	(1992)	(1992)	(1963)	(1982)	(1966)
	SMAX*k*P	69.0	105.1	115.1	123.3	150.1	167.0	–
		(1966)	(1996)	(1984)	(1984)	(1996)	(1969)	

Note: [1] closely followed by 105.6 mm (1995).

test period (1992). However, the *t*-test results indicate insignificance of both change points. The temporal change in AMAX30P in the basin appears to occur in 1991, but this is also not significant as indicated by the *t*-test result. Figure 4.8 shows the change point results of AMAX10P in the Meuse basin as well as in the two subcatchments.

The change point results of annual maximum *k*-day series for the two selected subcatchments are largely supportive. Most of the annual maximum *k*-day series (e.g. *k* = 5, 7, 10 and 15 days) appear to have notably increased since the beginning of the 1980s. Examples of AMAX10P in the two subcatchments are shown in Figure 4.8. However, AMAXP in the Geul subcatchment shows a significant decrease around 1973. No evident change was found for AMAXP in the Rur subcatchment. The inconsistency between the test results of AMAXP in the two subcatchments is likely due to the influence of local variability as mentioned earlier.

The majority of the annual maximum *k*-day series are subject to insignificant upward trends or do not show clear trends over the analysed periods, with the exception of AMAXP in the Geul subcatchment which shows an insignificant downward trend.

4.3.3 Winter precipitation extremes

Of all winter maximum *k*-day series, only WMAX10P in the Meuse basin for the short test period and WMAX15P in the Rur subcatchment show significant serial correlation for lag 1.

Winter maximum *k*-day series with *k* ≤ 10 for the Meuse basin show a significant

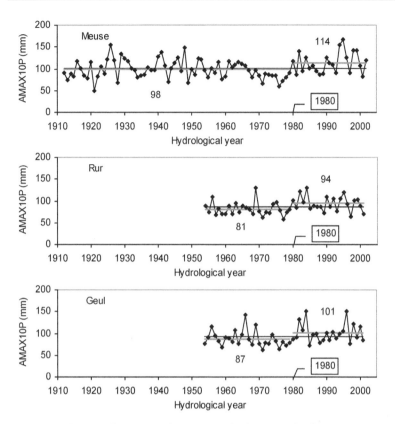

Figure 4.8 Change point results of AMAX10P in the Meuse basin (1912–2002) and the Rur
and Geul subcatchments (1954–2001).

increase most likely around the mid-1930s, without an evident change in the latter
sub-period. With increasing k (e.g. $k = 15$, …, 30 days), the increase became weak
and was not statistically discernible. Over the short test period, only WMAX7P in
the basin shows a significant increase around 1984.

Most of winter maximum k-day series for the two subcatchments show
significant increases occurring in the end of the 1970s or the beginning of the 1980s.
Figure 4.9 shows the change point results of WMAX10P in the Meuse basin and the
two subcatchments as examples.

The majority of the winter maximum k-day series are subject to upward trends,
of which only the trends for WMAX5P in the Meuse basin, and WMAXP and
WMAX5P in the Rur subcatchment are statistically significant.

4.3.4 Summer precipitation extremes

None of the summer maximum k-day series shows serial correlation for lag 1.

Summer maximum k-day series with $k \geq 5$ for the Meuse basin show a
significant increase around 1980 over the short test period, which became less
evident over the long test period. There seems to be little change at a shorter
duration (e.g. $k = 1$ and 3 days).

No obvious change was found in summer maximum k-day series for the two

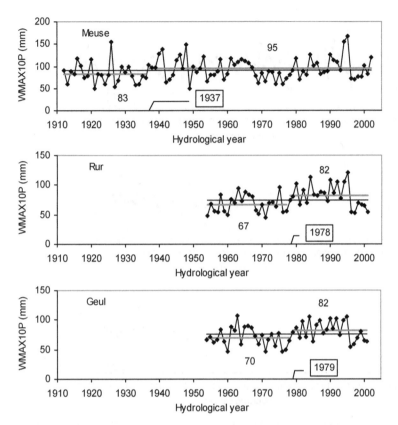

Figure 4.9   Change point results of WMAX10P in the Meuse basin (1912–2002) and the Rur and Geul subcatchments (1954–2001).

selected subcatchments except for SMAXP in the Geul subcatchment which shows a significant decrease since 1973.

None of the summer maximum *k*-day series is subject to a significant trend. Nevertheless, the downward trends in SMAXP in the two subcatchments are noticeable.

4.4 Change in the frequency of precipitation events

4.4.1 Analysed variables and test periods

Analyses of the frequency of occurrence of precipitation events in the Meuse basin are carried out based on the MeuseLP record (1911–2002) and the main results have been previously described in Tu *et al.* (2005b). Rainy days in the basin are defined as days with daily areal precipitation equal to or above 0.3 mm, since lower values measured at individual rain-gauging stations could be dew etc. The absolute thresholds of 1 mm/day and 10 mm/d are used to select wet days ($\geq$ 1 mm/d) and very wet days ($\geq$ 10 mm/d), respectively. The selected thresholds may be lower than those that are usually applied to a single station, but it should be realised that in this study averages of seven gauging stations are used.

Figure 4.10 depicts the distribution of monthly frequencies of rainy, wet and very wet days in the basin together with their associated precipitation amounts. The figure shows that the rainy days are almost evenly distributed during the year. Most days (about 81% on average) fall into the group of wet days. A few very wet days are observed for each month, accounting for nearly 14% of the rainy days in a year. For all months, the wet days are responsible for the major proportion (more than 95%) of the precipitation amount over the area, which implies that analysis of the wet days can be sufficient for elaborating the precipitation pattern change in the Meuse basin. The very wet days contribute almost half of the total precipitation amount of the rainy days. In this study, the frequency of occurrence of wet days (denoted as $WD_{\geq 1mm}$) and very wet days (denoted as $WD_{\geq 10mm}$) in the Meuse basin as well as their associated precipitation amounts (denoted as $PWD_{\geq 1mm}$ and $PWD_{\geq 10mm}$, respectively) are investigated. Apart from the annual scale (November–October), the winter (November–April) and the summer (May–October) are also considered. All derived series are examined within two time frames: 1912–2002 and 1950–2002.

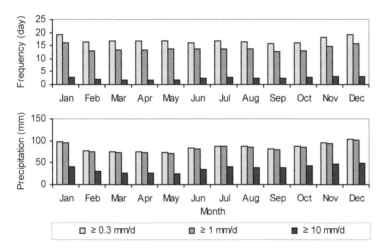

Figure 4.10 Monthly distributions of the defined precipitation events and their associated precipitation amounts in the Meuse basin (based on the MeuseLP, 1911–2002).

The very intense precipitation events such as heavy wet days ($\geq$ 20 mm/d) and extreme wet days ($\geq$ 30 mm/d) are not considered in the trend analyses due to limited occurrence of these events in the time series. Figure 4.11 shows the annual frequency series of both events in the Meuse basin. The occurrence of heavy wet days and extreme wet days in a year is rare, on average about 4.3 days and 0.9 days per year, respectively. Within the year, the occurrence of heavy to extreme wet days seems to be less in the spring months from March to May.

The frequency of occurrence of precipitation events in the Rur and Geul subcatchments is not analysed in this study. It is assumed that both subcatchments, being situated in the same climatic region, show similar long-term trends (to the entire Meuse basin), particularly in the winter time.

4.4.2 Frequency of annual precipitation events

The frequency of annual $WD_{\geq 1mm}$ in the Meuse basin appears to be relatively stable

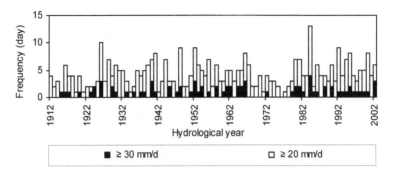

Figure 4.11 Time series of annual heavy wet days (≥ 20 mm/d) and extreme wet days (≥ 30 mm/d) in the Meuse basin (1912–2002). In the figure, the black bars are superimposed on the white bars.

in the 91 years under study. Nevertheless, the associated $PWD_{\geq 1mm}$ shows a pronounced increase (78 mm/a) since 1980. The average intensity of annual $WD_{\geq 1mm}$ has changed from 5.9 mm/d to 6.4 mm/d. Analysis of the frequency of annual $WD_{\geq 10mm}$ in the basin suggests a significant increase (by 5 day/period, about 19%), accordingly resulting in a significant increase in the associated $PWD_{\geq 10mm}$ (by 85 mm/period, about 20%) from 1980 onwards.

None of the annual frequency or amount series is subject to a significant trend, despite the different lengths of test periods. Over the 91 years, the upward trends in the frequency of annual $WD_{\geq 10mm}$ and the associated $PWD_{\geq 10mm}$ show high probabilities (> 90%).

4.4.3 Frequency of winter precipitation events

No obvious change was found for either the frequency of winter $WD_{\geq 1mm}$ or the resulting $PWD_{\geq 1mm}$ in the Meuse basin. Their average values for the study period are 86 day/period and 512 mm/period, respectively. However, there is evidence that the annual number of winter $WD_{\geq 10mm}$ in the basin has significantly increased (by 3 day/period) after 1980, leading to a simultaneous increase (by 57 mm/period) in the associated $PWD_{\geq 10mm}$ (see Figure 4.12). These two data series show a good linear relationship (r = 0.96). Furthermore, comparison is made for both $WD_{\geq 1mm}$ and $WD_{\geq 10mm}$ in the basin on the monthly scale before and after 1980 (see Figure 4.13). It appears that annually there have been a few more $WD_{\geq 1mm}$ (on average 2.6 day/period) in March after 1980, despite little change in the annual number of winter $WD_{\geq 1mm}$. The apparent increase in winter $WD_{\geq 10mm}$ in the basin after 1980 can be reasonably attributed to the increases of $WD_{\geq 10mm}$ occurring in the extended winter months from December to March.

No obvious trend was found for the number of winter $WD_{\geq 1mm}$ in the Meuse basin, neither for the associated $PWD_{\geq 1mm}$. The upward trends in the frequency of winter $WD_{\geq 10mm}$ and the resulting $PWD_{\geq 10mm}$ are not statistically significant.

4.4.4 Frequency of summer precipitation events

The annual number of summer $WD_{\geq 1mm}$ in the Meuse basin shows an insignificant decrease (on average 4 day/period) around 1969, without detectable change in the

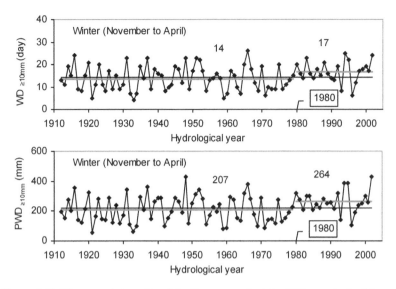

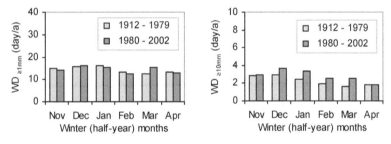

Figure 4.12 Change point results of the frequency of winter $WD_{\geq 10mm}$ in the Meuse basin and the associated $PWD_{\geq 10mm}$ (1912–2002).

Figure 4.13 Monthly distributions of winter $WD_{\geq 1mm}$ and $WD_{\geq 10mm}$ in the Meuse basin for the different periods: 1912–1979 and 1980–2002.

associated $PWD_{\geq 1mm}$. No obvious change was found for either the frequency of summer $WD_{\geq 10mm}$ or the associated $PWD_{\geq 10mm}$.

Except for an insignificant downward trend in the frequency of summer $WD_{\geq 1mm}$, no clear trend was observed for the summer $WD_{\geq 10mm}$ and the associated $PWD_{\geq 1mm}$ and $PWD_{\geq 10mm}$ in the Meuse basin.

4.5 Discussion

The statistical results obtained in sections 4.2 to 4.4 show some interesting changes in the precipitation regimes of the Meuse basin and the Rur and Geul subcatchments over the considered periods. The most remarkable change is the upward shift around 1980 for a number of related precipitation variables. There is a tendency to increased annual precipitation (APT) in the Meuse basin since 1980, accompanied by some varying changes in the seasons, e.g. a wetter spring (mainly March) from the late 1970s to the end of the 1980s and a slightly drier summer (mainly August) since the late 1960s. Additional analysis of the precipitation amount in extended winter (December–March) for the Meuse basin indicates a significant increase since 1979. The tendency to wetter conditions is in accordance with the fact that the intensity of (averaged) daily precipitation in the Meuse basin has notably increased after 1980, e.g. from 5.9 mm/day to 6.4 mm/day for wet days ($WD_{\geq 1mm}$) in the year. There are more frequent very wet days ($WD_{\geq 10mm}$) occurring in the winter half-year (particularly December–March) since 1980. This feature is likely (but not necessary) to have had effect on the precipitation extremes of certain duration (e.g. ranging from one day to 30 days) in the Meuse basin, although the quantitative relation between the two variables was not investigated. For example, most of the annual maximum consecutive k-day precipitation (AMAXkP) in the area show a simultaneous increase since 1980. The significant increase of winter maximum consecutive k-day precipitations (WMAXkP) with $k \leq 10$ in the Meuse basin since the mid-1930s (rather than 1980) was not expected. However, the evidence from the Rur and Geul subcatchments over the second half of the 20th century do support the likely effect on the winter multi-day precipitation extremes. The above qualitative description provides strong indications for systematic changes in the main features of the precipitation regimes in the Meuse basin during the study period.

Long precipitation time series might have been affected by instrumental changes and station move over the period of record. In section 2.4.1, quality control of the observed data has been performed by testing the annual precipitation series independently (without a reference series) for each station or through the double mass analysis. Moreover, it has been assumed that changes in the instrumentation and station location did not have a major impact on the discontinuities of the station records. Based on the available meta-data, there is no indication that the remarkable changes around 1980 for the analysed precipitation variables in this study are affected by instrumental and location changes. In fact, non-homogeneities of station records caused by instrumental and location changes are rarely systematic over an entire region. Therefore, those non-homogeneities and non-systematic errors at individual stations will tend to level each other out in the averaging of multiple station series for regional analyses (New *et al.*, 2001). However, one question may arise as to why non-homogeneities of the analysed precipitation series in this study appear evidently in the form of abrupt changes or shifts, instead of more gradual changes that are often assumed in climate change analyses. In the literature, there are also examples of shifts found in the precipitation data series, e.g. Tomozeiu *et al.* (2002) who found a significant downward shift around 1980 (at some stations)/1985 (at the rest of the stations) for the winter precipitation in Emilia-Romagna region (Italy), and Ho *et al.* (2003) who reported a sudden change around the late 1970s for the summer precipitation (more heavy precipitation) in Korea. Regarding the issue of abrupt climatic changes and oscillations, Lockwood (2001) reviewed of recent research work and stated that "the climatic system is viewed as a dissipative, highly non-linear system, under non-equilibrium conditions, and, as such, should be expected to have some unusual properties. These unusual properties include bifurcation points with marked instability just before the point, magnification of semi-periodic oscillations around bifurcation points, and variations in the strength of teleconnections with distance from equilibrium." From both the historical, Holocene and glacial climatic records, abrupt climatic changes and oscillations were found on all time-scales. Furthermore, the author stated: "the study of abrupt climatic changes is very new and raises many interesting and important questions about the way in which the climatic system operates. Climatic changes in the past have not always taken place in a slow, smooth manner. It is most unlikely that future changes associated with the present observed global warming will be smooth."

The analysis results obtained in this study demonstrate that there are indeed apparent abrupt changes in the precipitation regime in the Meuse basin as well as in the selected subcatchments, characterised essentially by enhanced annual precipitation and more intense precipitation events in the winter half-year since 1980. The findings are partly supportive to or in agreement with other studies for the Meuse basin and its vicinity. For example, KNMI (1999) and Pfister *et al.* (2000) both reported more winter precipitation in the second half of the past century for their analysed station series; Gellens (2000) found significant upward trends in winter multi-day extremes but none in summer multi-day extremes for the Belgian stations over the short period 1951–1995. It is also interesting to note the significant decrease (around 1973) identified for the annual daily extreme (AMAXP) in the Geul subcatchment. This tendency agrees with the finding of Vaes *et al.* (2002) that for small aggregation levels (i.e. less than one day) there is a small decrease in extreme precipitation events over the past century, which implies that the short and heavy thunderstorms certainly do not occur more frequently in recent years than before. The findings for the Meuse basin are also consistent with those reported for many parts in northwestern Europe, e.g. Schmith (2001) who found a positive trend in the winter (October–March) precipitation amount for most of 40 stations in northwestern Europe throughout the period 1900–1990, Klein Tank and Können (2003) who showed evidence of amplified response of precipitation extremes for the European stations where the annual precipitation amount increases in the period 1946–1999. Although regional variations are apparent, many studies have identified an increase in the "intensity" of daily precipitation over the 20th century in spite of differing definitions of intense precipitation (New *et al.*, 2001). The tendency to wetter conditions and more frequent intense events in the Meuse basin over the last couple of decades provides a good demonstration to the conclusion of Folland *et al.* (2001) that widespread increases are likely to have occurred in the proportion of total precipitation derived from heavy and extreme precipitation events over land in the mid- and high latitudes of the Northern Hemisphere.

4.6 Concluding remarks

Based on the statistical results and the discussion above, the following concluding remarks can be drawn:

- During the 20th century (1911–2002), the annual precipitation (APT) in the Meuse basin (upstream of Borgharen/Monsin) appears to have significantly increased since 1980, accompanied by some varying changes in the seasons, e.g. a wetter spring (mainly March) from the late 1970s to the end of the 1980s and a slightly drier summer (mainly August) since the late 1960s.
- The distribution of precipitation events in the Meuse basin has obviously changed since 1980, characterised essentially by more frequent very wet days ($WD_{\geq 10mm}$) in the winter half-year (particularly December–March) but without an equivalent change in the summer half-year (May–October). In other words, the winter wet days ($WD_{\geq 1mm}$) have become wetter over the recent two decades.
- The tendency to more intense precipitation events in the Meuse basin since 1980 is likely to have had effect on the precipitation extremes of multi-day duration (ranging from one day to 30 days), e.g. leading to a simultaneous

increase in most of the annual maximum consecutive k-day precipitation depth (AMAXkP) in the area. The significant increase in most of the winter maximum consecutive k-day precipitation depth (WMAXkP) in the basin since the mid-1930s seems to be affected little by the changing precipitation pattern since 1980, in view of noncoincidence of their change point years alone. However, the evidence from the selected subcatchments (Rur and Geul) over the second half of the 20th century do support the likely effect on the winter multi-day precipitation extremes.

Precipitation variations at a regional scale are generally associated with correspondingly large-scale atmospheric phenomena. In Chapter 5, an effort will be made to explore a possible linkage between the precipitation pattern change in the Meuse basin and the synoptic circulation patterns, intending to confirm the temporal changes identified for the selected precipitation variables and meanwhile to better understand the climate-induced precipitation variability in the Meuse basin. After Chapter 5, quantitative information of precipitation variability in the Meuse basin and the two selected subcatchments will be used for a discussion of the rainfall-runoff relationships in these areas and in turn for a proper interpretation of the temporal changes identified for the derived hydrological variables (see Chapter 6).

5   Linking precipitation variability to large-scale atmospheric circulation

5.1 Introduction

Atmospheric circulation, particularly in mid-latitudes, is the main control behind regional changes in temperature, precipitation and other climatic variables (Slonosky *et al.*, 2000). There is a growing interest in linking atmospheric circulation to local climate. Yarnal *et al.* (2001) and Tveito and Ustrul (2003) provided an extensive review of the development of classification systems and their application in synoptic climatology. Manual classification schemes have usually a long history and were developed for specific regions usually reflecting the dominating features of the atmospheric circulation influencing the climate in the targeted region. In Europe, the *Grosswetterlagen* system according to Hess and Brezowsky (see Gerstengarbe and Werner, 1999) for central Europe is probably the best known (Tveito and Ustrul, 2003). Objective classification systems were developed in the second half of the last century. Popular classification techniques are correlation-based analyses, eigenvector-based analyses (mainly two related techniques, principal components analysis and empirical orthogonal functions), composing and indexing (Yarnal *et al.*, 2001). The P–27 classification scheme developed at KNMI using the 500 hPa heights for nearly the whole of Europe (Kruizinga, 1979) is an example of the eigenvector-based classification systems. The North Atlantic Oscillation (NAO) index is the most well-known circulation index for Europe (Hurrell, 1995). Considering the geographical location of the Meuse basin in Europe and the long time period of concern, the *Grosswetterlagen* system and the NAO index are of most interest in this study. Brief descriptions of the *Grosswetterlagen* system and the NAO index were given in sections 2.6.1 and 2.6.2, respectively.

The prevailing westerly winds in mid-latitude Europe bring precipitation throughout most of the year. In the *Grosswetterlagen* system, the western circulation patterns are often used to establish the relation between precipitation variability and long-term variation of atmospheric circulation, particularly during winter months. Bárdossy and Caspary (1990) found that despite the nearly constant mean annual frequency for the period 1881–1989, the frequency of the western circulation patterns in the months of December and January has increased after 1973, while the frequency in the months of April and May has decreased after 1968; correspondingly, a decrease in the annual frequency of meridional "blocking" circulation patterns has been observed since 1980. Gerstengarbe and Werner (1999) also examined annual and seasonal frequencies of circulation patterns for the period 1881–1998 and confirmed changes in the frequencies of some circulation patterns. Apart from changes in the frequencies, changes in the persistence of circulation patterns (measured by the mean duration of circulation patterns) also received much attention. Bárdossy and Caspary (1990) noted that the transition frequencies for several different circulation patterns have changed markedly. Gerstengarbe and Werner (1999) and Werner *et al.* (2000) found that there was an increase in the duration of zonal circulation patterns in winter. Some studies have investigated the influence of circulation patterns using the *Grosswetterlagen* system on local precipitation. For example, Bouwer (2001) has demonstrated the impact of the zonal circulation patterns on the winter (DJF) precipitation amounts for nine stations in western Europe during the observation periods from the 1970s to the 1990s. Pfister *et al.* (2000) showed evidence of a marked increase in the contribution of the westerly airflows to precipitation in the Alzette basin (in the vicinity of the Meuse basin) since the 1950s, causing an increase in precipitation intensity and duration. Steinbrich *et al.* (2005) analysed the time series of the *Grosswetterlagen* and heavy precipitation (> 30 mm/d) for areas of about 440 km^2 and found clear spatio-temporal patterns and, thus, were able to relate heavy precipitation events to distinct groups of *Grosswetterlagen* in different areas of southwestern Germany. There are also some studies that focus on the link between floods and circulation patterns using the *Grosswetterlagen* system, e.g. Caspary (1995), Kaestner (1997), Steinbrich et al. (2002), Uhlenbrook et al. (2002). Steinbrich *et al.* (2002) and Uhlenbrook *et al.* (2002) demonstrated clearly the impact of western cyclonic and southwestern circulation patterns for the seasonality of floods in southwestern Germany. Some researchers have attempted to explore the link between droughts and circulation patterns. For example, Stahl and Demuth (1999) have identified circulation patterns associated with streamflow drought in the regions of southern Germany. Van de Griend (1981) mentioned that the *Grosswetterlagen* system appears to be less suitable for synoptic-climatological analysis in the Alpine region.

The NAO is a prominent mode of low-frequency variability of the Northern Hemisphere atmosphere and has its strongest effects in the winter season. The winter NAO shows large inter-annual and inter-decadal variability (Hurrell, 1995). Different versions of NAO indices are available via internet. Jones *et al.* (1997) remarked that the choice of the southern station can make some differences in seasons other than winter. The literature on variations in the NAO is numerous. Hurrell (1995) reported that from the turn of the past century until about 1930, there were strong positive anomalies in the winter (December–March) NAO index; from the early 1940s until the early 1970s, the winter NAO index was low and exhibited a downward trend; since then, a sharp reversal has occurred, with unprecedented strongly positive values since 1980. Moreover, Jones *et al.* (1997) described the late 1980s and early 1990s as the period with the highest values (strongest westerlies) and the winter of 1995–1996 as a dramatic switch in the winter (November–March) NAO index. Tomozeiu *et al.* (2002) detected an upward shift around 1980 for the winter (December–March) NAO index during the period 1960–1995. Precipitation variability was a major topic for studies using the indexing technique (Yarnal *et al.*, 2001). In Europe, roughly north of Paris a positive NAO generally results in large precipitation volumes and south of Paris there is a negative relation between the precipitation volume and the NAO index (Hurrel, 1995; Hurrell and Van Loon, 1997). In addition, Hurrell and Van Loon (1997) reported that wetter-than-normal conditions over northern Europe and Scandinavia since 1980 are linked to the behaviour of the NAO. Wibig (1999) found that the NAO mode has the greatest influence on precipitation in western Europe during the winter months (December–March). Some papers have investigated the influence of the NAO on precipitation in local areas in Europe. For example, Kotlarski *et al.* (2004) found clear statistical relationships between large-scale atmospheric flow conditions (parameterized by *Grosswetterlangen* and NAO index) and precipitation in southwestern Germany. These relationships are subject to a clear spatial variability and a pronounced temporal variability throughout the year. An enhanced zonality of atmospheric flow led to high amounts of precipitation especially in exposed areas (mountain ranges) and in wintertime. Tomozeiu *et al.* (2002) found that an intensification of the positive phase of the NAO especially after 1980 could be responsible for the decrease around 1985 in the winter (November–March) precipitation for 40 stations in Emilia-Romagna (northern Italy) during the period 1960–1995.

The objective of this chapter is to explore a possible linkage between precipitation variability in the Meuse basin and climate variability. Both the *Grosswetterlagen* system and the NAO index are used. The use of the *Grosswetterlagen* system is mainly restricted to the Meuse basin, with emphasis on explaining the precipitation pattern change in the area (section 5.2). The main analysis results have been reported in Tu *et al.* (2005b). The use of the NAO index extends to two selected subcatchments (Rur and Geul) in addition to the Meuse basin (section 5.3). Initially the (KNMI) P–27 classification scheme was also of interest to this study. However, the record according to this classification scheme is relatively short (1949–present) and is composed of two periods (i.e. before and after 1980) for which different analysis methods have been applied in the classification, thus likely introducing a potential inconsistency around 1980. The P–27 record is therefore not analysed in this study. A discussion of the statistical results is arranged in section 5.4 and then some concluding remarks are given in section 5.5.

5.2 *Grosswetterlagen* system

5.2.1 Influence of different weather types on the precipitation events

Based on the daily data of the *Grosswetterlagen* system for the period 1911–2002 (see section 2.6.1), Figure 5.1 depicts the monthly distribution of three (i.e. zonal, half-meridional and meridional) circulation groups and the major circulation patterns within the zonal circulation group. Along the year, the monthly frequencies of the zonal group (expressed as a percentage) are relatively low in April and May, which are balanced with the relatively high monthly frequencies of the meridional group in these two months. Among the zonal circulation patterns, the sub-type Wz (West cyclonic) dominates in all months.

In section 4.4, the number of wet days ($WD_{\geq 1mm}$) and very wet days ($WD_{\geq 10mm}$) in the Meuse basin was analysed. To identify the influence of different circulation patterns on the precipitation events, the percentages of three different circulation groups occurring on $WD_{\geq 1mm}$ and $WD_{\geq 10mm}$ only were computed. Figures 5.2 and 5.3 show the results of three circulation groups and different zonal circulation sub-types during $WD_{\geq 1mm}$ and $WD_{\geq 10mm}$, respectively. As seen from Figure 5.2, during the year, the zonal circulation patterns on $WD_{\geq 1mm}$ are nearly evenly distributed for most months, with slightly lower values for April and May. The relative contribution of the zonal circulations on $WD_{\geq 10mm}$ increases in terms of percentage from October to March. In Figure 5.3, the effect of the sub-type Wz on $WD_{\geq 1mm}$ and particularly on $WD_{\geq 10mm}$ is dominant. The monthly contributions from the zonal circulations to the precipitation on $WD_{\geq 1mm}$ range from 25% (May) to 50% (December). In the case of $WD_{\geq 10mm}$ the monthly contributions are a few to more than ten percent higher.

In the half-meridional and meridional circulation groups, a few circulation patterns, e.g. SWz, NWz, TM, Nz, HNz, TrM, Sz, TB and TrW (see Table 2.6), were also found to be important to the occurrence of $WD_{\geq 1mm}$ in the Meuse basin. Their individual chances of being $WD_{\geq 1mm}$ are generally more than 50%. Table 5.1

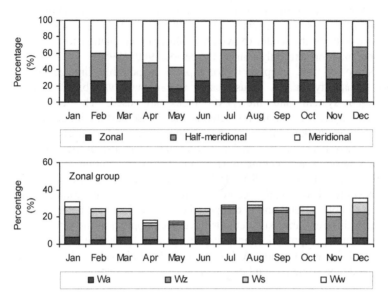

Figure 5.1 Percentages of three circulation groups and the major circulation patterns within the zonal group during the period 1911–2002. Notations: Wa for West anticyclonic, Wz for West cyclonic, Ws for Southern west and Ww for Angleformed west (see Table 2.6).

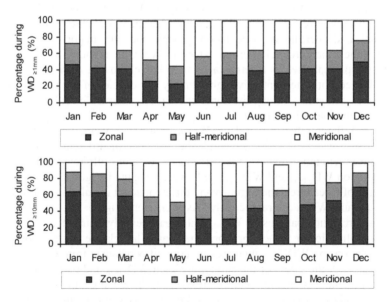

Figure 5.2 Percentage of three circulation groups that apply in the Meuse basin on wet days (WD$_{\geq 1mm}$) and very wet days (WD$_{\geq 10mm}$), respectively.

gives the frequencies of occurrence of these circulation patterns and their probabilities of being WD$_{\geq 1mm}$, respectively. Separately the contribution of these patterns to the frequency of WD$_{\geq 1mm}$ in the year is of minor importance, but their

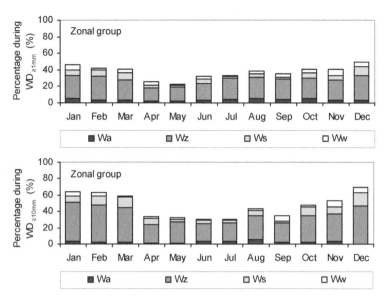

Figure 5.3 Percentages of four zonal circulation patterns that apply in the Meuse basin on WD$_{\geq 1mm}$ and WD$_{\geq 10mm}$, respectively.

combined contribution can be substantial, evidently during the months from April to September. Taking the sub-types SWz and NWz as examples, their resulting precipitation amounts in both WD$_{\geq 1mm}$ and WD$_{\geq 10mm}$ accounted for more than ten percent of the corresponding totals in most months. Particularly worthy of note is that some of these circulations may play an important role in the extreme events, especially from late spring to summer.

5.2.2 Analysed variables and test periods

In this study, the zonal circulation group and a few rain-associated circulation sub-types such as Wz, SWz and NWz are chosen to explore the links between the precipitation pattern change in the Meuse basin and climate variability. Their frequencies of occurrence (in percentage) and the resulting precipitation events (both WD$_{\geq 1mm}$ and WD$_{\geq 10mm}$) as well as the associated precipitation amounts are analysed on both annual (hydrological year) and seasonal (winter and summer half-years) scales. Detection of change points in the derived time series within two time frames, 1912–2002 and 1950–2002, is carried out by the Pettitt test only, because some time series include a substantial number of zero-values.

5.2.3 Temporal variability of rain-associated circulation patterns

The annual frequency of zonal circulation patterns has been relatively stable during the past century. However, the winter frequency has significantly increased (roughly by 5%) since 1972. The summer frequency shows an insignificant decrease (roughly by 3%) since 1968. The sub-type Wz and the sub-types SWz and NWz also show obvious changes in their frequencies. For example, the annual frequency of Wz and the annual frequency of SWz and NWz show a major increase around 1980 and 1948, respectively. Here the sub-types Wz, SWz and NWz are considered together.

Table 5.1 A few circulation patterns (CPs) with relatively high probabilities (in percentage) of being $WD_{\geq 1mm}$ in the Meuse basin during the period 1911–2002. The frequencies of occurrence (in percentage) of the CPs are given in brackets.

CP	Jan	Feb	Mar	Apr	May	Jun	Jul	Aug	Sep	Oct	Nov	Dec	Year
Half-meridional group													
SWz	71.4	68.1	68.0	62.7	66.7	82.1	64.7	61.8	85.3	74.1	76.5	76.1	71.9
	(4.9)	(2.8)	(1.8)	(2.4)	(3.1)	(1.4)	(1.8)	(1.9)	(2.7)	(4.1)	(3.6)	(3.1)	(2.8)
NWz	74.9	79.2	62.7	74.8	63.0	49.0	59.1	71.7	69.4	68.5	70.4	81.8	69.1
	(5.9)	(5.0)	(5.6)	(4.9)	(2.6)	(3.7)	(6.9)	(4.0)	(3.9)	(3.1)	(3.6)	(5.8)	(4.6)
TM	50.0	59.7	56.0	65.0	65.2	63.8	72.6	67.2	72.2	69.4	60.3	75.0	64.5
	(1.3)	(2.8)	(2.6)	(4.3)	(3.2)	(2.1)	(2.2)	(2.0)	(2.0)	(1.7)	(2.5)	(1.4)	(2.3)
Meridional group													
Nz	75.3	66.7	63.0	64.3	47.7	65.3	60.0	63.2	73.1	78.7	58.1	69.1	63.9
	(2.6)	(2.4)	(3.2)	(4.1)	(4.6)	(3.6)	(1.6)	(2.0)	(2.4)	(1.6)	(2.7)	(2.4)	(2.8)
HNz	61.4	55.8	67.3	46.6	58.0	50.8	61.4	57.1	44.4	46.2	38.9	43.5	54.6
	(1.5)	(1.7)	(1.9)	(2.1)	(3.5)	(2.1)	(1.5)	(1.0)	(0.7)	(1.4)	(0.7)	(0.8)	(1.6)
TrM	62.0	65.2	76.0	65.0	58.3	70.4	75.3	62.8	64.5	67.7	66.3	61.0	66.5
	(3.8)	(3.5)	(3.4)	(6.5)	(2.9)	(4.2)	(5.1)	(3.3)	(4.5)	(3.4)	(6.4)	(3.5)	(4.2)
Sz	45.7	59.3	47.4	45.0	—	100	—	50.0	66.7	62.5	65.5	67.3	59.7
	(1.2)	(2.1)	(0.7)	(0.7)		(0.1)		(0.1)	(0.4)	(1.1)	(2.1)	(1.9)	(0.9)
TB	81.8	77.1	76.9	73.5	67.9	75.5	73.4	66.7	71.9	56.1	75.4	75.9	71.4
	(1.2)	(1.8)	(1.4)	(2.5)	(4.6)	(1.8)	(3.3)	(4.2)	(2.1)	(2.0)	(2.1)	(1.9)	(2.4)
TrW	51.2	56.1	60.6	53.4	69.9	71.4	77.0	63.2	71.5	71.1	65.6	54.9	65.5
	(1.4)	(1.6)	(2.5)	(6.3)	(5.8)	(4.8)	(4.4)	(4.8)	(4.5)	(3.2)	(4.5)	(1.8)	(3.8)

Note: No occurrence of the sub-type Sz in May and July.

Both annual and winter frequencies of the combined sub-types show a significant increase around 1960, while the summer frequency appears to be nearly constant since the early decade of the last century. Figure 5.4 depicts the change point result of the winter frequency for the long test period. In the figure, the winter frequency of the combined sub-types Wz, SWz and NWz shows a significant increase of 8% from 1960 onwards as compared with the corresponding mean prior to 1960. The increase since 1960 concentrated on the months from November to March. Over the short test period starting from 1950, the significant increases in the annual and winter frequencies occurred most likely around 1980 or 1981. The winter frequencies for the sub-periods 1950–1980 and 1981–2002 are about 24% and 30% (on average), respectively.

5.2.4 Linking precipitation variability to rain-associated circulation patterns

The $WD_{\geq 1mm}$ and $WD_{\geq 10mm}$ in the Meuse basin due to the zonal circulation patterns did not show significant changes in their annual frequencies. Neither did the

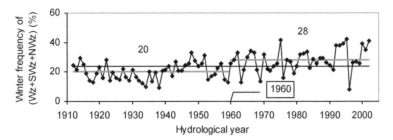

Figure 5.4 Change point result of the winter (November–April) frequency of the combined sub-types Wz, SWz and NWz (Wz+SWz+NWz) over the period 1912–2002.

associated precipitation amounts. However, the winter and summer frequencies of the associated $WD_{\geq 1mm}$ have noticeably changed since 1970 and 1968, respectively. Their resulting precipitation amounts show a significant increase around 1979 and a slight decrease around 1949, respectively. The winter frequency of the associated $WD_{\geq 10mm}$ and the corresponding precipitation amount show an insignificant increase after 1945. Regarding the Wz-associated $WD_{\geq 1mm}$ and $WD_{\geq 10mm}$ in the Meuse basin, their annual frequencies were found to have significantly increased since 1980. The change in the winter frequency of the Wz-associated $WD_{\geq 10mm}$ as well as the resulting precipitation amount is characterised by a distinct step-wise increase, first around 1940 and then around 1980.

This study explores whether the frequencies of the combined sub-types Wz, SWz and NWz can be linked to the precipitation pattern change in the Meuse basin. The annual frequencies of the associated $WD_{\geq 1mm}$ and $WD_{\geq 10mm}$ in the basin have significantly increased from 1980 onwards, causing concurrent changes in their resulting precipitation amounts. Similar results were obtained for the winter half-year. For the summer half-year, there was also an (insignificant) increase in the associated $WD_{\geq 10mm}$ after 1980, but without an evident change in the associated $WD_{\geq 1mm}$. Figure 5.5 illustrates the change point results of the annual frequency of the associated $WD_{\geq 1mm}$ and the resulting precipitation amount. After 1980, the annual frequency of the $WD_{\geq 1mm}$ in the basin contributed by the combined sub-types Wz, SWz and NWz shows an increase from 15% to 19%, causing an increase of 144 mm/period (about 38%) in the associated precipitation amount. The observed increase in the associated $WD_{\geq 1mm}$ (and also $WD_{\geq 10mm}$) after 1980 is mainly due to the monthly changes from September to March. Figure 5.6 illustrates the change point results of the winter frequency of the associated $WD_{\geq 10mm}$ and the resulting precipitation amounts. After 1980, annually there was a significant increase from 4% to 6% (about 4 day/period) in the winter frequency of the associated $WD_{\geq 10mm}$, accompanied by a significant increase of 78 mm/period (about 73%) in the resulting precipitation amount. Accordingly the average intensity of winter $WD_{\geq 10mm}$ in the basin has increased from 15 mm/d to nearly 17 mm/d.

The tendency to increased $WD_{\geq 1mm}$ (and $WD_{\geq 10mm}$) in the Meuse basin due to the combined sub-types Wz, SWz and NWz in the winter half-year reflects to a large extent the increasing trend in the winter frequency of these rain-associated circulation patterns. A rather high correlation ($r = 0.95$) was obtained between these two frequency series. It was also found that out of the winter $WD_{\geq 1mm}$ in the Meuse basin (see section 5.4), the relative percentage of $WD_{\geq 1mm}$ due to the combined sub-types Wz, SWz and NWz has obviously increased. A similar trend was also

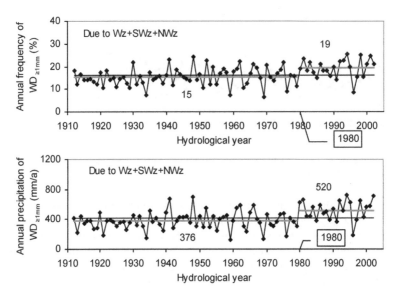

Figure 5.5 Change point results of the annual frequency of $WD_{\geq 1mm}$ in the Meuse basin due to the combined sub-types Wz, SWz and NWz (Wz+SWz+NWz) and the associated precipitation amount (1912–2002).

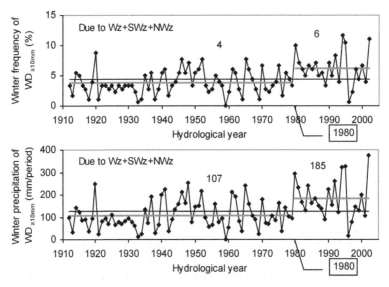

Figure 5.6 Change point results of the winter frequency of $WD_{\geq 10mm}$ in the Meuse basin due to the combined sub-types Wz, SWz and NWz (Wz+SWz+NWz) and the associated precipitation amount (1912–2002).

observed in the relative percentage of the associated $WD_{\geq 10mm}$ in the Meuse basin. Furthermore, Figure 5.7 compares the precipitation amount in the Meuse basin with the frequency of the combined sub-types Wz, NWz and SWz for the winter and summer half-years, respectively, based on the 10-year moving averages. In the

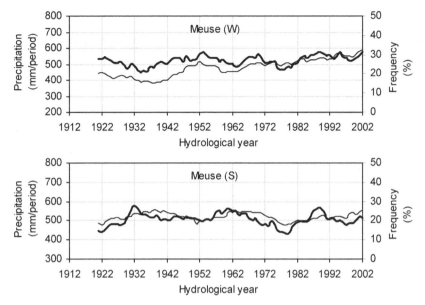

Figure 5.7 Comparison of the precipitation amounts (the left x-axis, thick lines) in the Meuse basin with the frequencies of the combined rain-associated circulation patterns Wz, NWz and SWz (the right axis y, thin lines) for the winter (W) and summer (S) half-years, respectively, based on the 10-year moving averages (1912–2002).

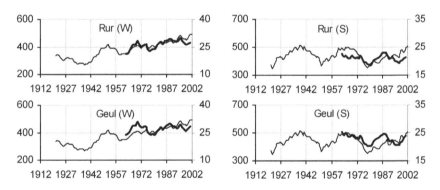

Figure 5.8 The same as in Figure 5.7 except that the Rur and Geul subcatchments (1954–2001) are analysed.

winter half-year, both time series show upward trends over the recent decades, although decadal and shorter cycles are also obvious in the time series. The results of the Spearman test indicate significance of the upward trends (based on the 10-year moving averages) over the period 1960–2002. It appears that the relationship between the two series has systematically altered from the 1960s onwards, which coincides with the change in the circulation patterns. In the summer half-year, the relationship appears to be relatively stable. Similar comparisons were also made for the Rur and Geul subcatchments (see Figure 5.8) and a few rain gauging stations in the vicinity of the Meuse basin (see Figure 5.9). In the figures, the winter

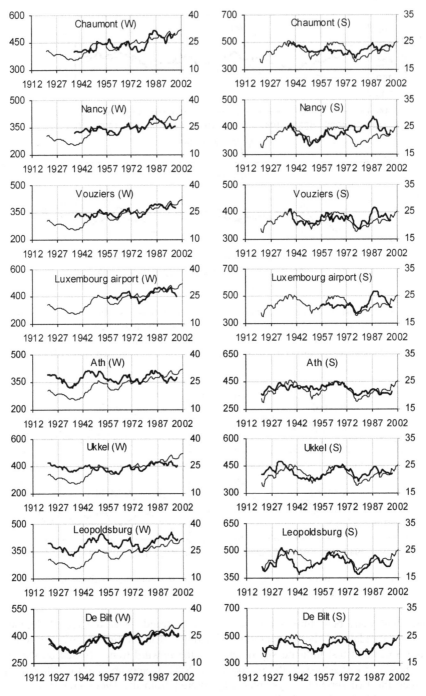

Figure 5.9 Comparison of the precipitation amounts (the left x-axis, thick lines) at nine rain gauging stations (with different record lengths) in the Meuse basin and its vicinity with the frequencies of occurrence of the combined rain-associated circulation patterns Wz, NWz and SWz (the right axis y, thin lines) for the winter (W) and summer (S) half-years, respectively.

precipitation amounts in the two subcatchments and at most stations indeed present increasing trends since the 1960s, roughly in accordance with the trend of the winter frequency of the typical rain-associated circulation patterns. The greatest consistency was observed at De Bilt. The results for the surrounding stations highlight the existence of strong statistical links between the precipitation variability (in the winter half-year) in the Meuse basin and the fluctuation of large-scale weather patterns over the recent decades.

5.3 North Atlantic Oscillation (NAO) index

5.3.1 Analysed variables and test periods

The winter (DJF) NAO index of the Gibraltar-minus-Iceland version (see section 2.6.2) is used to make a link with WINP in the Meuse basin and the two selected subcatchments (Rur and Geul). The winter NAO index values are obtained by averaging the monthly values of the constituent months (December–February). Detection of change points in the winter (DJF averaged) NAO index series is performed for the entire study period 1912–2002. Further segmentation is considered. The relation between WINP in the study areas and the winter NAO index is examined using correlation analysis.

5.3.2 Temporal variability of the winter NAO index

No significant serial correlation for lag 1 was found for the winter NAO index series.

Over the entire study period 1912–2002, the Pettitt test detected a major change point around 1938 for the winter NAO index series. Further segmentation indicates a secondary change point around 1979 for the sub-period 1939–2002 and a subsequent change point around 1962 for the sub-period 1939–1979. However, the statistical significance of these change points in the Pettitt test results (with probabilities of 0.88, 0.88 and 0.84, respectively) is not very high (below 0.90). Figure 5.10 presents the fluctuation pattern of the winter NAO index series. The average value of the winter NAO index from 1939 to 1962 is close to the long-term mean of 0.53. The shift from the negative average value to the high positive average value round 1980 is very striking. However, compared with the sub-period prior to 1939, the last couple of decades is not exceptional. Summarily, the predominantly positive phase of the NAO during the winter half-years in the sub-periods prior to 1939 and after 1980 reflects the enhanced surface westerlies over Europe.

5.3.3 Correlation between winter precipitation and winter NAO index

Figure 5.11 compares the time series of WINP in the Meuse basin and the winter NAO index for the entire period of record. A visual analysis of these records indicates that there is roughly an agreement between their fluctuation patterns except for notable disparities or discrepancies during the period prior to the mid-1930s, the first half of the 1940s, the first half of the 1970s and the period from the late 1980s to the early 1990s.

Correlation analyses of WINP in the study areas and the winter NAO index were carried out for the different periods. Table 5.2 presents the correlation coefficients.

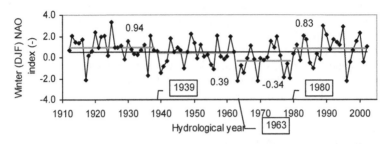

Figure 5.10 Change point results of the winter (DJF averaged) NAO index over the period 1912–2002.

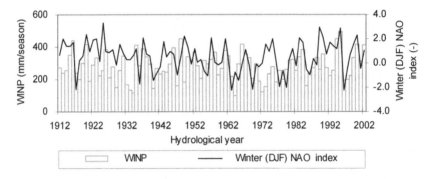

Figure 5.11 Comparison of WINP in the Meuse basin with winter (DJF averaged) NAO index (1912–2002).

Table 5.2 Correlation coefficients (r) between WINP in the study areas and the winter (DJF averaged) NAO index for the different periods.

Catchment	1912–2002/ 1954–2001[1]	1912–1938	1939–1962	1963–1979	1980–2002/ 1980–2001[1]
Meuse	**0.40**	0.38[2]	0.34	0.10	**0.59**
Rur	**0.52**	–	–	0.21	**0.66**
Geul	**0.47**	–	–	0.18	**0.62**

Note: [1] For the Rur and the Geul; [2] improved to **0.63** after excluding the "outliers" in the years of 1917, 1925 and 1936. All bold values are significant at a 1% level.

Over the 91-year period, WINP in the Meuse basin shows a significant positive correlation (at a 1% significance level) with the winter NAO index, especially after 1980. During the sub-period 1963–1979, there was on average a low value for the winter NAO index. Moreover, the correlation with WINP was relatively poor for this period. The r value of 0.38 for the sub-period 1912–1939 is not as high as expected (just significant at a 5% level), seemingly due to the influence of the "outliers" occurring in the years of 1917, 1925 and 1936. After excluding the "outliers", the linear correlation improved markedly, with the r value of 0.63

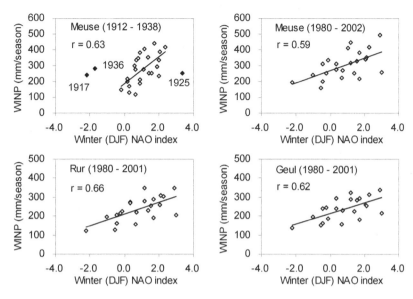

Figure 5.12 Scatter plots of WINP in the study areas against winter (DJF averaged) NAO index for the periods of high positive NAO index average values. Black points represent the "outliers" occurring in the years of 1917, 1925 and 1936 are not used in the correlation analysis. Linear regression lines calculated by least squares fitting are added.

Table 5.3 Correlation coefficients (r) between WINP at individual stations used for the Meuse basin and the winter NAO index for the different periods.

Period	Maredsous	Rochefort	Thimister	Stavelot	Hives	Chimay	Chiny
1912–2002	**0.34**	**0.34**	**0.40**	**0.42**	**0.39**	**0.39**	**0.36**
1912–1938[1]	**0.56**	**0.52**	**0.66**	**0.65**	**0.56**	**0.54**	**0.62**
1980–2002	**0.54**	**0.59**	**0.61**	**0.66**	**0.59**	**0.55**	0.46

Note: [1] After excluding the "outliers" occurring in the years of 1917, 1925 and 1936. All bold values are significant at a 1% level.

(significant at a 1% level). Figure 5.12 shows the scatter plots of WINP in the Meuse basin against the winter NAO index for the two sub-periods 1912–1939 and 1980–2002. Similar relations with the winter NAO index were found for WINP in the Rur and Geul subcatchments during the second half of the 20th century. Figure 5.12 also shows their scatter plots for the sub-period after 1980 for illustration. For the common periods 1954–2001 and 1980–2001, the r values (0.52 and 0.66, respectively) for the Rur subcatchment appear to be slightly higher than the r values for the Geul subcatchment (0.47 and 0.62, respectively) and for the Meuse basin (0.46 and 0.60, respectively).

Correlation analysis was also extended to individual stations used for the Meuse basin. Table 5.3 lists their correlation coefficients for the entire period (1912–2002) and the two sub-periods (1912–1938 and 1980–2002). The r values for all stations are significant at a 1% level, except for the r value for Chiny during the sub-period

1980–2002 being significant at a 5% level. Higher correlation with the winter NAO index was found at Thimister and Stavelot, which are located in the northern part of the Meuse basin.

5.4 Discussion

The statistical results in section 5.2 demonstrate that the prevailing westerly circulation patterns over Europe, described by the *Grosswetterlagen* system, have changed considerably at least during the most recent two decades. The tendency to more intense precipitation events (e.g. increased $WD_{\geq 10mm}$) and subsequently larger precipitation amounts in the Meuse basin since 1980 is very likely linked to the increase in the frequency of occurrence of the rain-associated circulation patterns (e.g. Wz, SWz and NWz), particularly in the winter half-year. These results are roughly in accordance with the previously published studies by Bárdossy and Caspary (1990), Pfister *et al.* (2000) and Bouwer (2001). Pfister *et al.* (2000) concluded that winter (October–March) precipitation variability in the Alzette basin is mainly due to fluctuations in the atmospheric circulation patterns. The Meuse basin and the Alzette basin are neighbouring basins and are both located in the transition zone between northwestern Europe with a wet and temperate climate and southern Europe with a Mediterranean climate, characterised by low annual precipitation. In this study, the circulation patterns on the previous days (e.g. two days) prior to rainy days are not analysed. In the literature, an increase in the duration of zonal circulation patterns in winter has been reported (e.g. Gerstengarbe and Werner, 1999; Werner *et al.*, 2000). Steinbrich *et al.* (2002) and Uhlenbrook *et al.* (2002) did not found a big difference when analysing the circulation patterns on the previous one to three days prior to the event.

   The statistical results in section 5.3 demonstrate that the surface westerlies across the North Atlantic into Europe during the winters (DJF), described by the winter (DJF) NAO index, has greatly strengthened since 1980. Correlation analysis of WINP in the Meuse basin and the winter NAO index indicates a significant positive correlation (at a 1% level) for the period 1912–2002. This evidence is supported by a similar correlation analysis for the Rur and Geul subcatchments. Compared with the general relations between precipitation and NAO in Europe (i.e. positive correlation over northern Europe and negative correlation over southern Europe) reported by Hurrel (1995) and Hurrell and Van Loon (1997), the correlation found for the Meuse basin appears to be relatively large compared to the findings of other studies. Schmith (2001) also reported evidence of the positive correlation during the winters (October–March) for nearly 40 stations in northwestern Europe throughout the period 1900–1990, in which seven Belgian stations used for the Meuse basin in this study were included.

   In this study, it was also found that the monthly precipitation totals for March in the Meuse basin and the two selected subcatchments were not clearly correlated with the NAO index for March that appears to have an upward shift around 1976 and then return to the mean level from the end 1990s onwards. Van Oldenborgh *et al.* (2000) reported a strong connection between strong warm El-Niño winter (DJF) events, characterised by the winter (DJF) $NINO_3$ index (a common measure of the strength of El-Niño), and high spring precipitation in a band from southern England eastwards into Asia. The possible mechanisms of this teleconnection were explained by southeast Asian surface temperature anomalies acting as intermediate variables. Based on locally measured temperature and wind direction at De Bilt, Van

Oldenborgh and Van Ulden (2003) found that there was an increase in southwesterlies (fewer northeasterlies) during the months of February to April after 1950.

With the (limited) investigations conducted in this study, it is not possible to make the causes clear of the changing synoptic patterns during the last couple of decades of the 20th century. Analysis of the changing climate system is beyond the scope of this study. A possible explanation for the abrupt change identified in the rain-associated circulation patterns using the *Grosswetterlagen* system may be linked to the NAO that appears to show decadal (and longer time-scale) fluctuations. Hurrell (1995) demonstrated that since the early 1980s, the NAO has remained in one extreme phase during the winter, contributing significantly to the recent wintertime warmth across Europe. An evaluation of atmospheric moisture budget reveals coherent large-scale changes since 1980 that are linked to the recent wetter conditions in northern Europe. However, the causes for such variability in the Atlantic are not clear and the relation of the NAO to greenhouse gas forcing is not completely understood yet.

5.5 Concluding remarks

Based on the statistical results and the discussion above, the following concluding remarks can be drawn:

- During the 20th century, the precipitation pattern change in the Meuse basin since 1980 is very likely linked to the fluctuation of large-scale atmospheric circulation, as characterised by the *Grosswetterlagen* system. The annual (and the winter half-year) frequencies of wet days ($WD_{\geq 1mm}$) and very wet days ($WD_{\geq 10mm}$) in the basin contributed by the rain-associated circulation patterns (e.g. Wz, SWz and NWz) have increased considerably since 1980.
- The winter precipitation (WINP) in the Meuse basin and the selected subcatchments (Rur and Geul) show a positive correlation (significant at 1% level) with the winter (DJF) NAO index during the entire periods under study. The tendency to more abundant precipitation in the Meuse basin and the subcatchments in winter roughly since 1980 (in the last several decades) is likely to be a consequence of the strengthened NAO that brings stronger westerly winds across the North Atlantic into Europe.

The synoptic-climatological analysis carried out in this chapter improves our understanding of the precipitation variability in the Meuse basin. The apparent changes identified in the precipitation events/amounts and extremes are likely to be climate-related. This general conclusion provides a scientific basis for subsequent analyses of the rainfall-runoff relationships in the study areas (Chapter 6).

6 Assessment of the rainfall-runoff relationships

6.1 Introduction

Runoff is defined here as the part of precipitation that appears as streamflow. The generation processes are widely described in literature (e.g. Newson, 1995; Brooks *et al.*, 1997; Bonell, 1998; Beven, 2001b; Uhlenbrook, 2005). Precipitation is viewed as the major input to a catchment and a key to its water yield characteristics. Besides precipitation, evapotranspiration is in general the most significant hydrological factor that determines runoff generation. The dynamics of hydrological processes are driven by meteorological factors, but the changes that these factors cause for the discharge regime do not occur instantaneously (WMO, 1994). The non-linear and time-variant nature of the relation between precipitation and runoff makes it difficult determining and quantifying changes in the relationship. Moreover, the derivation of the rainfall-runoff relationship over a catchment is very much dependent on the time scale (being considered) and is also affected by the spatial scale being considered (Shaw, 1991; Blöschl and Sivapalan, 1995).

In the previous Chapters 3 and 4, the "observed" changes in the discharge and precipitation regimes in the Meuse basin as well as in the selected subcatchments have been identified by statistically analysing a great number of hydrological and hydro-meteorological variables. This chapter makes an assessment of the historical rainfall-runoff relationships in the study areas (emphasizing the entire Meuse basin). The potential impacts of human intervention will be specifically evaluated in Chapter 7. Chapter 6 is organised as follows: Section 6.2 focuses on the general rainfall-runoff relations. Long-term water balance analyses are carried out first and then the temporal changes in the rainfall-runoff relations are investigated. Section 6.3 explores the links between generated floods and antecedent precipitation in the study areas. Section 6.4 illustrates the effects of precipitation and evapotranspiration on the low flows. After the largely qualitative analysis of the rainfall-runoff relationships based on the statistical results in the forementioned sections, section 6.5 provides a hydrological modelling study to gain further insights into the rainfall-runoff relationship in the entire Meuse basin. A discussion of the statistical results and the simulated results is made in section 6.6. Finally, some concluding remarks are drawn in section 6.7.

6.2 Changes in the rainfall-runoff relations

6.2.1 Long-term water balance analyses

The components of the water balance in a catchment (for which the surface and groundwater divides coincide) generally include precipitation P, evaporation E (soil evaporation, interception and transpiration are considering collectively and the last item is dominating here), runoff R, and change in storage (e.g. surface water, soil moisture and groundwater etc.) ΔS. The water balance equation, which applies for a specific period Δt, may be written as:

$$P - E - R = \Delta S / \Delta t \qquad (6.1)$$

The dimension of the components in Eq. 6.1 is depth per time step [L.T^{-1}]. There are various methods (e.g. commonly Arithmetic mean method, Thiessen polygon method and Isohyetal method) to estimate the areal precipitation (P) from the values observed by the rain gauges in or close to the catchment. The runoff (R) is obtained from streamflow measurements at the outlet of the catchment. The change in the amount of water stored in the catchment (ΔS) predominantly takes place in the subsurface (soil moisture and groundwater) in humid temperate climate if longer time scales (e.g. more than some days) are considered. The component ΔS may be large for small periods (e.g. day or month), but is often negligible for long-term averages, in particular if the year is carefully selected (hydrological year or water year). When the values in Eq. 6.1 represent long-term annual averages, the assumption that ΔS = 0 can often be accepted and the annual average of E from the catchment can then be calculated as the difference between P and R (WMO, 1994).

Using Eq. 6.1 and assuming no change in storage, long-term water balance analyses for the Meuse basin (upstream of Monsin) and the selected subcatchments (Semois, Rur and Geul) were carried out not only over their entire study periods but also over the sub-periods which were segmented at the years of very suspicious non-homogeneities in the discharge records (see Chapter 3). Table 6.1 summarizes the results of these analyses plus the values of runoff coefficient (RC, defined as the percentage of precipitation that produces to runoff). In the table, the precipitation record for Chiny (located in the middle part of the area, see Figure 2.4) was used for the Semois subcatchment. Compared with the Semois areal precipitation record (generated by KMI from data of a dense network in the area), the Chiny station record appears to overestimate the annual precipitation by a few percent (nearly 50 mm/a) over the common period 1968–1998. Both records show a fairly good correlation on an annual basis ($r = 0.95$). Nevertheless, nothing is known about their relation in the earlier years prior to 1968. In Table 6.1, the mean annual values of PET$_{P-M}$ for De Bilt (see section 2.5.1) were used to represent the maximum possible loss rates under the prevailing meteorological conditions in the entire Meuse basin. However, one has to realise that the actual ET is affected by the soil moisture available. Moreover, the estimates of potential ET apply for grass and do not reflect the real situation of historical land use in the catchment area, and the station De Bilt is located outside the Meuse basin. Therefore, the computed PET$_{P-M}$ values are associated with considerable uncertainties, if related to the study areas, and should not be simply compared with the values of water loss (WL), which are computed as the difference between annual averages of P and R over the considered period. The item WL includes not only ET, ΔS and catchment leakage, but also measurement errors (Newson, 1995). In Table 6.1, the standard deviation for each item, computed from the annual series, is provided to indicate the degree of variability and uncertainty in the associated mean annual value.

The long-term average runoff of a basin depends on the physio-geographical conditions (Sokolovskii, 1971). In the Meuse basin, the headwater areas in the Ardennes produce more runoff due to generally higher precipitation, less permeable soils and relatively steep slopes, as illustrated by the example of the hilly Semois

Table 6.1 Mean annual values and standard deviations (given in brackets) for the hydrological components in the Meuse basin and the selected subcatchments.

Catchment	Period	P[1] (mm/a)	R (mm/a)	RC (%)	WL (mm/a)	PET[2] (mm/a)
Meuse	1912–2000	949 (157)	406 (114)	42 (8)	543 (93)	562 (38)
(near Monsin)	1912–1932	948 (164)	441 (104)	46 (6)	506 (92)	556 (28)
	1933–2000	950 (157)	395 (115)	41 (8)	555 (91)	564 (41)
Semois	1930–2002	1278 (236)	710 (197)	55 (9)	568 (128)	564 (40)
(at Membre)	1930–1967	1287 (238)	726 (186)	56 (8)	561 (131)	564 (42)
	1968–2002	1269 (237)	693 (210)	54 (10)	577 (126)	563 (39)
Rur (at Stah)	1954–2001	836 (131)	303 (71)	36 (6)	533 (93)	556 (41)
Geul	1954–2001	900 (143)	334 (90)	37 (7)	567 (105)	556 (41)
(at Meerssen)	1954–1970	923 (153)	367 (110)	39 (9)	556 (104)	544 (41)
	1971–2001	888 (138)	316 (72)	36 (7)	572 (106)	562 (40)

Note: RC = 100·(R/P); WL = P-R. [1] The MeuseRLP record for the Meuse; the Chiny station record for the Semois; the Geul7P record for the Geul; [2] based on the PET_{P-M} record for De Bilt.

subcatchment. In Table 6.1, the mean annual RC value (over the period 1930–2002) for the Semois is more than 10% higher than the mean annual RC value (over the period 1912–2000) for the Meuse. The runoff proportions for the Rur and the Geul appear to be less, nearly in the same order (36 to 37%) over the latter half of the 20th century.

As indicated in Table 6.1, the mean annual WL values for the Meuse basin and the selected subcatchments over their entire study periods are relatively similar (on average 550 mm/a) and approximate the mean annual PET_{P-M} values for De Bilt. Nevertheless, these absolute values should be viewed as coarse estimates only. The Jeker subcatchment is not included in Table 6.1 because no precipitation data are available. Moreover, as being located in the chalky aquifer, the effect of river-groundwater interaction (i.e. deep seepage) in this area can not be ignored in water balance analysis.

Figures 6.1 and 6.2 depict the monthly distributions of the basic water balance components using bar graphs, in which 12 months are shown in the sequence of a hydrological year instead of a calendar year. In Figure 6.1, the monthly distributions for the Meuse and the Semois over their entire study periods are similar, generally characterised by higher runoff values during winter (November–April) and lower runoff values during summer (May–October). Application of the shorter period for instance 1954–2001 for both rivers makes no difference. The runoff distributions for the Rur and the Geul show more balanced patterns, primarily as the consequences of the reservoir regulation (in the Rur) or the relatively stable baseflow (in the Geul). In Figures 6.2, the water losses (WL) in the study areas are largely governed by the changing PET_{P-M} rate and the change in subsurface storage. Normally, the groundwater system gains recharge in the winter season and releases water in the summer to sustain the dry-season flow.

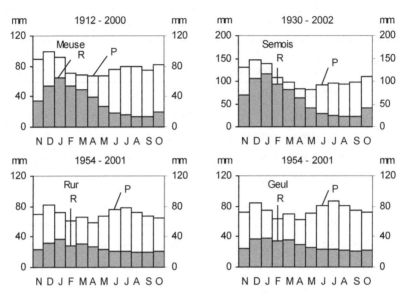

Figure 6.1 Monthly distribution of precipitation (P, mm/month) and runoff (R, mm/month) in the Meuse basin (upstream of Monsin) and the selected subcatchments Semois (upstream of Membre), Rur (upstream of Stah) and Geul (upstream of Meerssen). The letters N, D, J, F, M, A, M, J, J, A, S, O represent 12 months in the sequence of a hydrological year (November–October). The same definitions of 12 months apply to the subsequent figures.

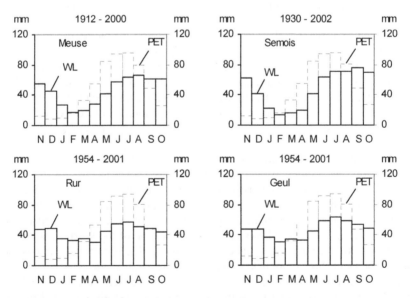

Figure 6.2 Monthly distribution of water losses (WL, mm/month) in the Meuse basin and the selected subcatchments plus potential evapotranspiration (PET$_{P-M}$) for De Bilt. WL = P-R.

6.2.2 Analysed variables and test periods

Continuous discharge and precipitation records used for Table 6.1 are used to derive time series of two lumped hydrological variables for the study areas (excluding the Jeker): annual runoff/rainfall ratio (i.e. annual runoff coefficient) and annual water loss. By considering the hydrological year, the changes in storage (mainly subsurface) in the catchment areas are in general minor, as reflected in their scatter plots of annual precipitation and annual runoff (with statistically significant correlation coefficients). Certainly this assumption is not valid for those very dry years which could cause soil moisture deficit for more years. As ET affects the soil moisture condition of a catchment and, consequently, affects the runoff process, the meteorological variable of PET_{P-M} for De Bilt on an annual basis is also examined.

Quarter-flow (or half-flow) interval, defined as the shortest time (typically in days) during which one quarter (or one half) of the annual runoff volume occurs, has been reported to be especially sensitive to land use changes (e.g. Black, 1991). Any change in the soil infiltration capacity or in one or more of the major storage components of the catchment can be expected to produce a (significant) change in the flow interval. Moreover, this parameter seems to be more consistent from year to year and within hydrographic regions. Analysis of the quarter-flow intervals for the study areas is therefore added. The unit of the flow interval is a unit of time in days, counted from the beginning of each hydrological year, i.e. 1 November.

The test periods of the derived variables are the same as the study periods defined in Table 6.1, except for the test period of annual PET_{P-M} rate for De Bilt which covers the whole period of record (1912–2002). The test period for the Jeker spans from 1965/1966 to 2001. Emphasis is placed on detection of change points in the time series. Sections 6.2.3 to 6.2.6 present the test results.

The variable of annual runoff/rainfall ratio describes the hydrological response of a catchment in a global sense and may not be sufficient to reveal the changes that may have taken place within the year. However, evaluation of the rainfall-runoff relations on a seasonal basis is difficult due to the effect of natural storage processes (mainly driven by ET) in the soil. Moreover, other factors such as human intervention and inconsistencies of the observation records complicate the interpretation. A synthesized evaluation of the seasonal changes in the rainfall-runoff relations will be included in section 7.2.

6.2.3 Change in the annual runoff/rainfall ratios

Only the annual runoff/rainfall ratio series for the Geul shows significant serial correlation for lag 1.

The annual runoff/rainfall ratio for the Meuse shows a significant decrease since 1932. The probability of the change point year 1931 in the Pettitt test result is high (0.94). A secondary change around 1977 was also detected. However, the t-test result indicates that the increase of the ratio from 1978 onwards was not large enough to make it statistically significant over the analysed period 1932-2000. Figure 6.3 shows the change point result of the ratio series for the Meuse. The mean values of runoff proportion before and after 1932 are 0.47 and 0.41, respectively.

No significant changes or trends were found for the ratio series of the selected subcatchments (see Figure 6.3), indicating that the annual rainfall-runoff relations in these areas remain relatively stable during their study periods. In Figure 6.3, the ratio series for the Rur and the Geul vary with time in a similar way except for a

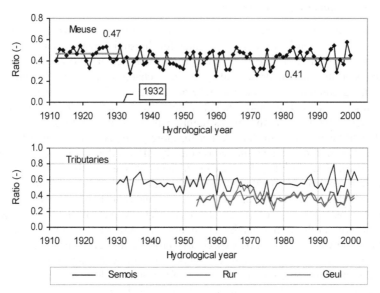

Figure 6.3 Change point result of annual runoff/rainfall ratio for the Meuse basin (upstream of Monsin, 1912–2000), and time series of annual runoff/rainfall ratios for the selected subcatchments Semois (upstream of Membre, 1930–2002), Rur (upstream of Stah, 1954–2001) and Geul (upstream of Meerssen, 1954–2001).

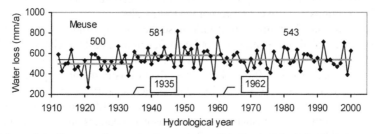

Figure 6.4 Change point result of annual water loss in the Meuse basin (upstream of Monsin, 1912–2000).

questionable short-term inconsistency shown for the Geul (roughly from the mid-1960s to the mid-1970s). A comparison with APT (annual precipitation) in the Rur subcatchment suggests that APT in the Geul subcatchment during that specific period was slightly underestimated, thus leading to those higher ratio values.

6.2.4 Change in the annual water losses

None of the time series of annual water loss for the Meuse basin and the selected subcatchments shows significant serial correlation for lag 1.

The test result of annual water loss in the Meuse basin suggests a notable increase since 1935, particularly from 1935 to 1961 (see Figure 6.4). Compared with the mean annual value of 500 mm/a over the earlier year prior to 1935, the mean annual values over the latter two sub-periods, 1935–1961 and 1962–2000, appear to have increased by 81 mm/a and 43 mm/a, respectively. No significant change points could be detected in the time series of annual water loss in the subcatchments.

6.2.5 Fluctuation of potential evapotranspiration

In the correlogram, the r_1 value (0.19) of the time series of annual PET_{P-M} rate for De Bilt is close to the upper 95% confidence limit (0.21).

As shown in Figure 6.5, the fluctuation of annual PET_{P-M} rate for De Bilt is in general characterised by relatively high rates from 1928 to 1950 and from 1989 onwards, and relatively low rates prior to 1928 and between 1951 and 1989. The difference between the high and low rates is in the order of some 30 mm/a. Additional trend analysis of the PET_{P-M} rate in the combined months of March to August, during which water loss in the area is predominantly due to PET (see Figure 6.2), reveals a similar fluctuation pattern (not shown), whereas the PET_{P-M} rate in the combined months of September to February has changed little over time.

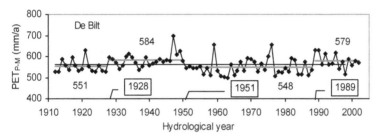

Figure 6.5   Change point result of annual PET_{P-M} rate for De Bilt (1912–2002).

6.2.6 Change in the quarter-flow intervals

Only the time series of quarter-flow interval for the Semois shows significant serial correlation for lag 1 ($r_1 = -0.34$, beyond the lower 95% confidence limit of -0.23).

No significant change point was detected for the quarter-flow interval of the Meuse, neither for the quarter-flow intervals of the selected tributaries (see Figure 6.6). This suggests that changes in soil moisture storage at the beginning of the hydrological year are very limited. The mean values of the interval durations are all less than three months over their study periods. A few years have a longer duration, e.g. 1922 (112 days) and 1954 (124 days). It is worth noting that the quarter-flow intervals of the Meuse and the Semois behave similarly.

6.3 Effect of antecedent precipitation on the flood peaks

6.3.1 Correlation between flood peak discharge and antecedent precipitation depth

Precipitation is considered the major driving force for flood generation in the Meuse basin. Antecedent catchment wetness and antecedent precipitation depth are two important factors influencing flood magnitudes and volumes. In the winter half-year, the soil moisture condition in large parts of the entire Meuse basin usually tends to be very wet at the beginning of a flood and therefore ordinary storm events might well cause significant peaks. In the summer half-year, because of generally large moisture deficit in the soil, intense precipitation events (often occurring locally in space) might not cause notable peaks at large scale. In this study, the winter maximum (WM) series (used interchangeably with WMAXD in section 3.3) of the

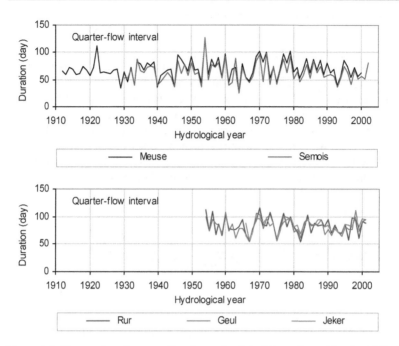

Figure 6.6   Time series of quarter-flow intervals of the Meuse (near Monsin, 1912–2000) and
the selected tributaries Semois (at Membre, 1930–2002), Rur (at Stah, 1954–
2001), Geul (at Meerssen, 1954–2001) and Jeker (at Nekum, 1966–2001). The
time of quarter-flow interval is counted from the beginning of each hydrological
year, i.e. 1 November.

Meuse (at Borgharen) and the selected tributaries Semois, Rur and Geul are used to
explore the correlation with their antecedent k-day precipitation depths (denoted as
AkP-WM, where k = 1, 3, 5, 7, 10, 15 and 30 days) which are computed by
summing the daily precipitations on the day of the flood and the previous k-1 days.

Table 6.2 gives the correlation coefficients for both the entire periods of record
and the sub-periods which are segmented around the change point years in the WM
series (i.e. WMAXD). Figure 6.7 shows part of the scatter plots over the entire
periods of record for illustration. Over the 91-year period, the peak discharges of
WM events in the Meuse appear to correlate closely with the antecedent 5 to 15-day
precipitation depths on wet soils (e.g. r = 0.77 for the duration of seven days). For
the Semois, application of the Chiny station record for the 73-year period produces
higher correlation coefficients for the duration of seven to ten days (e.g. r = 0.61 for
the duration of seven days). Such a defined duration for the Rur appears to be
longer, covering 15 to 30 days, which is probably related to the impact of reservoir
regulation on the river (see section 7.3). For the Geul, the influences of the
measurement system change (after 1971) and the flood mitigating measures need to
be considered. The duration corresponding to higher correlation coefficients after
1971 appears to have shifted from a day or a few days to a couple of weeks,
inferring that the flood mitigating measures might be largely responsible for the
dominating decrease in WMAXD of the Geul, because a measurement system
change would hardly cause a notable change in the defined duration. Another factor
obscuring the correlation for the Geul is groundwater flow. The peak of fast runoff

Table 6.2 Correlation coefficients between the peak discharges of WM events and their antecedent k-day precipitation depths for the Meuse and the selected tributaries.

River	Period	AkP-WM						
		$k=1$	$k=3$	$k=5$	$k=7$	$k=10$	$k=15$	$k=30$
Meuse[1]	1912–2002	**0.29**	**0.58**	**0.71**	**0.77**	**0.74**	**0.73**	**0.66**
(at Borgharen)	1912–1983	0.24	**0.55**	**0.74**	**0.77**	**0.71**	**0.70**	**0.62**
	1984–2002	0.40	0.44	**0.57**	**0.68**	**0.72**	**0.76**	**0.77**
Semois[2]	1930–2002	0.24	**0.57**	**0.57**	**0.61**	**0.61**	**0.53**	**0.40**
(at Membre)	1930–1978	0.04	**0.62**	**0.67**	**0.69**	**0.64**	**0.60**	**0.57**
	1979–2002	**0.45**	**0.71**	**0.65**	**0.68**	**0.64**	**0.60**	**0.45**
Rur	1954–2001	0.27	0.17	0.32	**0.43**	**0.54**	**0.70**	**0.76**
(at Stah)	1954–1979	0.27	0.10	0.37	0.47	0.47	**0.58**	**0.78**
	1980–2001	0.46	0.10	0.18	0.27	**0.52**	**0.76**	**0.72**
Geul[3]	1954–2001	**0.39**	**0.52**	**0.48**	**0.44**	**0.50**	**0.54**	**0.60**
(at Meerssen)	1954–1970	**0.65**	**0.74**	**0.69**	0.55	0.55	0.49	0.56
	1971–2001	0.37	0.31	**0.47**	**0.46**	**0.58**	**0.70**	**0.66**

Note: [1] The MeuseLP record is used. [2] The Chiny station record is used. [3] The Geul5P record is used. All bold values are significant at a 1% level.

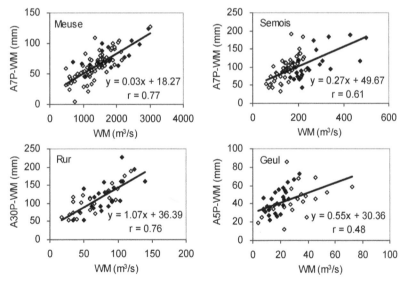

Figure 6.7 Scatter plots of the magnitudes of WM against the antecedent k-day precipitation depths (AkP-WM) for the Meuse (at Borgharen, 1912–2002) and for the selected tributaries Semois (at Membre, 1930–2002), Rur (at Stah, 1954–2001) and Geul (at Meerssen, 1954–2001), respectively. Black points represent the floods occurring after the change point years (see Table 6.2). Linear regression lines for all scatter points are added in the figures.

in conjunction with the maximum base flow could lead to a severe flood in the Geul (Dautrebande et al., 2000).

The flood-producing processes are complex. Analysis of antecedent precipitation in the catchment can provide only rough cause-effect results of the flood regime of a river. A major difficulty in such studies has been to account for effects of precipitation intensities. Besides, the orientation (i.e. the direction of movement) of storm events in the area may have an important influence on the resultant floods (WMO, 1994). Other natural factors such as snow coverage or snowmelt and frozen soil are also important to flooding on some occasions (Shaw, 1991). These influential factors are not considered in this study.

6.3.2 Analysed variables and test periods

Section 6.3.1 demonstrates that the magnitudes of WM events in the Meuse as well as in the selected tributaries (Semois, Rur and Geul) are in general linearly and positively correlated (significant at a 1% level) with their corresponding antecedent precipitation depths over several days or weeks. In the subsequent sections 6.3.3 to 6.3.4, the study areas of interest are confined to the Meuse basin and the Semois subcatchment. The following variables are analysed in order to see whether the relation between peak discharge and precipitation has changed over time:

i) the antecedent 10-day precipitation depths preceding the WM/WPOT event (denoted as A10P-WM and A10P-WPOT, respectively) in the Meuse, the antecedent 7-day precipitation depths preceding the WM/WPOT event (denoted as A7P-WM and A7P-WPOT, respectively) in the Semois; and

ii) the ratios (used as a crude indicator to filter out the influence of precipitation variability in time) of the WM/WPOT peak discharge and A10P-WM/WPOT for the Meuse, the ratios of the WM/WPOT peak discharge and A7P-WM/WPOT for the Semois.

During the antecedent ten or seven days, the preceding 3rd and 4th day in general have more intense daily precipitations for all cases. The precipitation depths of these two days account for more than 40% of A10P-WM/WPOT for the Meuse and for more than 50% of A7P-WM/WPOT for the Semois. Considering the importance of the preceding 3rd and 4th day for both rivers, the precipitation depths cumulated over these two days (denoted as $P_{3rd+4th}$-WM/WPOT) are added for trend analysis.

The derived variables are examined over the entire study periods. Further segmentation is also considered to allow more detailed insights.

6.3.3 Change in the antecedent precipitation depths

The WM events

Analysis of the A10P-WM series for the Meuse shows a significant increase around 1984 over the 91-year period, coupled by a distinct short-term decrease from 1969 to 1983 (see Figure 6.8). This fluctuation pattern is rather similar to that shown in the WMAXD series for Borgharen (see section 3.3.2). Compared with the earlier sub-period 1912–1968, the mean value of A10P-WM for the Meuse after 1984 has increased by 16 mm (about 20%). No significant change point in A7P-WM for the Semois was detected.

Antecedent precipitation depths over longer durations, e.g. ten days, 15 days and 30 days, were checked for both rivers. The test results suggest no obvious change. However, analysis of the $P_{3rd+4th}$-WM series for the Meuse reveals a consistent fluctuation pattern with that of the A10P-WM series. Relative to the sub-period prior

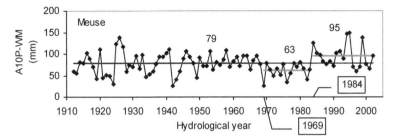

Figure 6.8   Change point result of A10P-WM for the Meuse (at Borgharen, 1912–2002).

to 1968, the percentage of increase in the $P_{3rd+4th}$-WM series after 1984 is around 24%. The above change is not evident in the $P_{3rd+4th}$-WM series for the Semois, which shows a change point year around 1978 with a low probability of 0.62 in the Pettitt test result.

The WPOT events

Segmentation of the A10P-WPOT$_{800}$ series for the Meuse produces a similar fluctuation pattern that is shown in the magnitude of the WPOT$_{800}$ events (see section 3.4.4). However, no significant change was detected for the A10P-WPOT$_{1217}$ series and the A10P-WPOT$_{1500}$ series for the Meuse. The A10P-WPOT$_{800-1500}$ series shows a change point located on 16 January 1943, which is just statistically significant (with a probability of 0.80) according to the Pettitt test. Further, the t-test result indicates insignificance of the change. Figure 6.9 shows the change point results of the A10P-WPOT series for the Meuse.

For the Semois, segmentations of the A7P-WPOT$_{90}$ series and the A7P-WPOT$_{90-180}$ series reveal a short-term rise from 1958 to 1968, followed by a decrease since 1968. The tendency towards decreasing was also observed in both the A7P-WPOT$_{158}$ series (around 1970) and the A7P-WPOT$_{180}$ series (around 1985). Figure 6.10 illustrates the change point results of the A7P-WPOT series for the Semois.

Analyses of the antecedent precipitation depths of longer durations (e.g. ten days, 15 days and 30 days) and the $P_{3rd+4th}$-WPOT for the two rivers lead to similar findings.

6.3.4 Change in the peak discharge-precipitation ratios

The WM events

No significant change was found in the ratio of magnitude of the WM events and A10P-WM for the Meuse (see Figure 6.11). The similarly defined ratio series for the Semois shows a clear step-wise increase, first around 1970 and then around 1982.

The WPOT events

The peak discharge/precipitation ratio series for both WPOT$_{800}$ and WPOT$_{800-1500}$ in the Meuse are nearly stable over the 91-year period, while the ratio series for WPOT$_{1217}$ and WPOT$_{1500}$ in the river show an insignificant increase (indicated by the t-test results) around 1951 or 1956. The probabilities of the change points for the two ratio series according to the Pettitt test are 0.95 (located on 12 February, 1950)

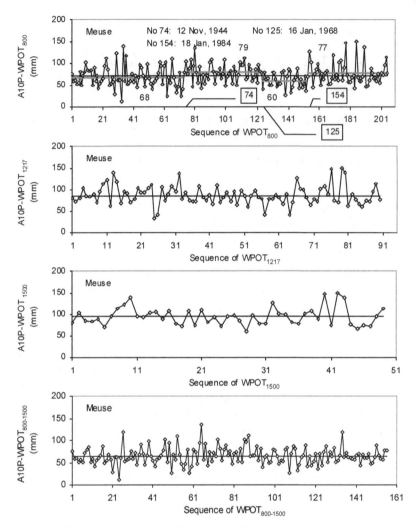

Figure 6.9 Change point results of A10P-WPOT for the Meuse (at Borgharen, 1912–2002).

and 0.89 (located on 22 December, 1952), respectively. All ratio series for the Semois show a significant increase around 1968 or 1970, with a further increase (around 1985) detectable in the ratio series for $WPOT_{180}$ in the river.

6.3.5 Contribution of antecedent precipitation on flood runoff

The trend results for the Meuse obtained in sections 6.3.3 to 6.3.4 imply that the increase of winter peak discharges over the last couple of decades (see sections 3.3.2 and 3.4.4) largely corresponds to the temporal variability of antecedent 10-day precipitation depth over the catchment area. However, the trend results obtained for the Semois appear to be somewhat suspicious. The A7P-WM/WPOT series (excluding $A7P\text{-}WPOT_{180}$) for the river show a distinct downward shift around

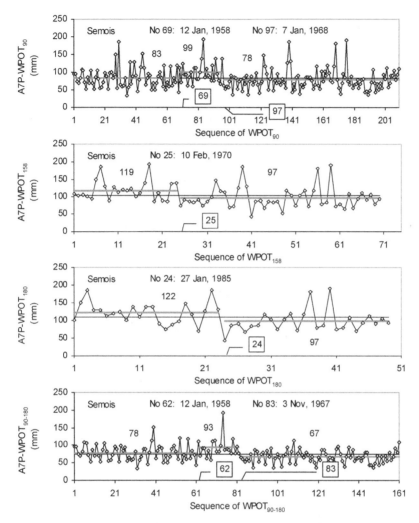

Figure 6.10 Change point results of A7P-WPOT for the Semois (at Membre, 1930–2002).

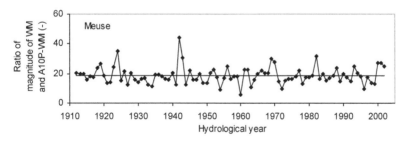

Figure 6.11 Change point result of the ratio of WM and A10P-WM for the Meuse (at Borgharen, 1912–2002).

1967/1968 or 1970, leading to a concurrent but upward shift in all peak discharge/precipitation ratio series. The non-homogeneity coincides with the measurement system change (around 1967/1968) of the Semois (at Membre). However, the latter factor appears to have no significant effect on the magnitudes of floods (see Figures 3.8 and 3.14) in the river. Therefore, a large doubt arises about the application of the Chiny precipitation record for the Semois subcatchment on a daily basis. Moreover, for the Semois subcatchment, A7P-WM/WPOT is perhaps an unsuitable measure relating to the peak discharge because snow coverage or frozen soil occasionally took place in this mountainous area, which can increase flood runoff generation. Comparatively, the peak discharges in the Meuse are assumed to be less sensitive to such natural factors due to its large spatial scale. In the following paragraphs, evaluation of the contribution of antecedent precipitation on flood runoff is confined to the Meuse basin.

In statistical analysis, the well-known double mass technique is perhaps the most frequently applied method for testing the relative consistency of the observations against another data set. Whenever a break or change point is found in the double mass curve, the non-homogeneous subset of the test variable can be further corrected, e.g. by following the procedure described in Allen *et al.* (1998). The success of the double mass analysis depends upon the test and reference variables being in constant ratio. The complex forms of relationship between precipitation and streamflow are liable to generate spurious breaks in double mass curves (Hall, 2001). In this study, it appears that the considerable inter-annual variations of the peak discharge series and the associated antecedent precipitation series do not justify correction of the non-homogeneity of the peak discharge series using the double mass method.

To gain insight into the relative changes, the average percentage changes in the magnitudes of winter floods and the associated antecedent precipitation depths after the potential change point years (including those statistically insignificant changes as indicated by the t-test results) have been computed (see Table 6.3). In the table, the winter peak discharges for the specified sub-periods mostly show the same direction of change as the antecedent precipitation depths. It is recognised that the mean values of the two variables for floods of different sizes in the Meuse actually contain overlapping information due to partly sharing common floods and therefore are not independent. Nevertheless, with the assumption that the winter peak discharges in general increase (non-linearly) with the enhanced associated A10P-WM/WPOT, any significant change in the historical relations would be more or less reflected in the mean values of winter peak discharges over the specified sub-period. In this study, simple linear regression analyses of the mean values for the early (reference) sub-periods and for the latter sub-periods for the Meuse were carried out separately, and then the established regression equations were used to provide coarse estimates of the relative changes in the observed magnitudes of winter flood events in the river during the second half of the 20th century. Figure 6.12 shows the linear regression lines plus the estimated regression equations and the relative changes in percentage. In the figure, the dominating influence of antecedent precipitation on the peak discharge of the Meuse becomes clear. Nevertheless, there seems an indication that during the second half of the 20th century, the winter medium and large floods in the Meuse have a tendency of being slightly greater. Take the winter floods of 800 m³/s, 1217 m³/s, 1500 m³/s (i.e. three POT thresholds) in the present situation as examples. Their percentage changes (relative to the first half of the 20th century) are some -5%, +3% and +6%, respectively. However, one should be aware of considerable uncertainty in the generalised percentage changes (as illustrated in

Table 6.3 Mean values of the winter peak discharges (WM/WPOT) in the Meuse (at Borgharen) and mean values of the associated antecedent 10-day precipitation depths (A10P-WM/WPOT) in the basin for the different sub-periods/periods.

Flood event	Sub-period/ period	WM/WPOT		A10P-WM/WPOT	
		m^3/s	Change (%)	mm	Change (%)
WM	1912–1970	1402		79	
	1971–1983	1052	-25	63	-20
	1984–2002	1771	+26	95	+20
$WPOT_{800}$	1912–1944	1196		68	
	1944–1968	1371	+15	78	+15
	1968–1984	1061	-11	61	-10
	1984–2002	1392	+16	77	+13
$WPOT_{1217}$	1912–1951	1570		90	
	1951–1984	1553	-1	78	-13
	1984–2002	1799	+15	91	+1
$WPOT_{1500}$	1912–1956	1785		96	
	1956–1993	1800	+0.8	90	-6
	1993–2002	2107	+18	100	+4
$WPOT_{800-1500}$	1912–2002	1087		65	

Note: Segments in shade are taken as the reference sub-periods/periods. The sign "+" indicates a relative increase; the sign "-" indicates a relative decrease.

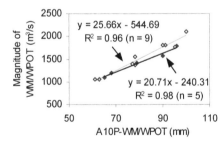 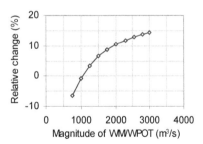

Figure 6.12 Scatter plots (left) of the mean values of the magnitudes (y) of WM/WPOT events in the Meuse (at Borgharen) against the mean values of A10P-WM/WPOT (x) for the different sub-periods/periods (see Table 6.3), and the relative percentage changes (right) in the observed magnitudes of winter floods during the second half of the 20th century (roughly corresponding to the latter sub-periods). Black points ($n = 5$) represent the data set from the reference sub-periods/periods, while blank points ($n = 9$) represent the data set from the latter sub-periods. The linear regression lines plus the regression equations are added.

Figure 6.12), primarily arising from the assumption of a linear relation between peak discharge and precipitation, the highly dependent data used for regression, the relatively large inter-annual variations of the two variables and the neglectance of other precipitation characteristics (e.g. rain intensity). Moreover, the fluctuation of ET could to some extent affect the soil moisture conditions preceding the floods in the early and late winter (e.g. November, March and April). The floods during these months are in general moderate or small and thus are more sensitive to the change in the moisture conditions. Therefore, the computed percentage values should be viewed largely as qualitative rather than quantitative estimates. An improved quantitative estimate of the relative change in the peak discharge could be analysed through a process-based modelling approach (e.g. Beven, 2001b).

6.4 Effects of precipitation and evapotranspiration on the low flows

6.4.1 Introduction

The groundwater recharge, often linked to dry-season flows, might be increased or decreased as a result of precipitation variability and, consequently, could influence the magnitude of low flows (Stahl, 2001). In humid and temperate areas, precipitation surpluses favour groundwater recharge in, normally, the winter season (Nonner, 2003). Besides precipitation, evapotranspiration is another main meteorological factor affecting low flows. Several studies have addressed the topic of low flows or droughts (i.e. extreme low flow situations) in the Meuse and its tributaries. For example, Uijlenhoet *et al.* (2001) carried out a multi-site hydrological drought study and found that the discharge fluctuations of the Meuse and its tributaries during their observation periods (1968–1998) appear to be related with the lithological/geological characteristics of the catchment areas. However, this study did not combine the observed discharge data with precipitation and evaporation data sets. De Wit *et al.* (2001) made a connection between the summer baseflow (calculated as the average value of the monthly minimum discharges for May to October) index and the winter (November–April)/summer (May–October) precipitation indices (formulated as Eq. 6.3, see section 6.4.2) for the Meuse (1911–1998). They found that the summer baseflow index shows some correlation to both winter and summer precipitation indices.

Aim of this sub-section is to explore the physical links between the changes in the summer low flows of the Meuse and the selected tributaries and the changes in meteorological conditions such as precipitation and evapotranspiration. Graphical analysis and correlation analysis are applied to determine the degree of association.

6.4.2 Relation with variations of precipitation

The relation between precipitation and summer low flows was examined by correlating the summer baseflow index (denoted as IBF_{summer}) with the precipitation index (denoted as IP_{period}) for the specified time periods including preceding winter half-year, current summer half-year and the hydrological year.

The equation for constructing IBF_{summer} is formulated as follows:

$$IBF_{summer,i} = BF_{summer,i} / BF_{summer,avg} \qquad (6.2)$$

where BF_{summer} is the summer baseflow (i.e. SMIN10D, m^3/s), i is the specific (hydrological) year and *avg* is the average value of SMIN10D for the entire period of record. A baseflow index value smaller than 1 suggests a lower-than-average baseflow, whereas a baseflow index value larger than 1 suggests a higher-than-average baseflow.

Similarly, IP_{period} is computed using the following formula:

$$IP_{period,i} = P_{period,i} / P_{period,avg} \tag{6.3}$$

where P_{period} is the precipitation depth cumulated during the specified time period, i is the specific (hydrological) year, and avg is the average value for the entire period of record. A precipitation index value smaller than 1 represents a dry period, whereas a precipitation index value larger than 1 represents a wet period.

Table 6.4 summarizes the correlation coefficients between IBF_{summer} and IP_{period} for the Meuse and the selected tributaries. In the table, not only the entire period of record, but also the sub-periods (segmented around the major change point years of SMIN10D) were considered. Correlation coefficients between between IP_{winter} and IP_{summer} are also given in the table to illustrate that IP_{winter} and IP_{summer} are not significant correlated. For all specified time periods, IBF_{summer} of the Meuse (near Monsin) appears to correlate closer with IP_{summer} than IP_{winter}, and best with IP_{annual}. Likewise for IBF_{summer} of the Semois. Different results were obtained for the Rur and the Geul. The correlation results for the Rur are not consistent with time. For the entire period of record, IBF_{summer} of the Rur tends to correlate stronger with IP_{winter}. IBF_{summer} of the Geul appears to be associated with IP_{winter} to a higher degree.

Table 6.4 Correlation coefficients (r) between IBF_{summer} and IP_{period} and between IP_{winter} and IP_{summer} for the Meuse and the selected tributaries.

River	Period	IBF_{summer} and IP_{period}			IP_{winter} and IP_{summer}
		IP_{winter}	IP_{summer}	IP_{annual}	
Meuse	1912–2000	**0.44 (0.47)**	**0.58 (0.55)**	**0.65 (0.67)**	0.17
(near Monsin)	1912–1932	0.51	**0.68**	**0.78**	0.17
	1933–2000	**0.49 (0.58)**	**0.60 (0.58)**	**0.71 (0.75)**	0.18
Semois[1]	1930–2002	0.28 (0.30)	**0.57 (0.55)**	**0.53 (0.53)**	0.16
(at Membre)	1930–1970	0.32	**0.67**	**0.63**	0.12
	1971–2002	0.38 **(0.64)**	**0.57 (0.50)**	**0.59 (0.72)**	0.21
Rur	1954–2001	**0.55**	0.20	**0.53**	0.02
(at Stah)	1954–1964	0.49	0.21	0.48	0.13
	1965–1987	0.35	**0.57**	**0.56**	0.37
	1988–2001	**0.74**	0.09	0.58	0.27
Geul	1954–2001	**0.56**	0.34	**0.60**	0.09
(at Meerssen)	1954–1988	**0.67**	0.37	**0.66**	0.19
	1989–2001	**0.67**	0.12	0.59	0.15

Note: [1] The Chiny precipitation record is used. The coefficients in brackets are computed after excluding the year of 1987 in which the unusually high SMIN10D values are found for the Meuse and the Semois. All bold values are significant at a 1% level.

Further, Figure 6.13 illustrates the evolutions of IBF_{summer} and IP_{period} for the Meuse and the Semois, revealing more details of the relations between these two index variables. IBF_{summer} of the Meuse prior to the mid-1930s (after moving five years) shows a clear upward shift relative to the latter period and appears to vary in a consistent way with IP_{summer}. After the mid-1930s (excluding some ten years from the mid-1950s to the mid-1960s), IBF_{summer} of the Meuse tends to be linked to IP_{winter} of the Meuse basin. It appears that the wetter conditions in the previous winter could have had an influence on SMIN10D of the Meuse. For the short 10-year period noted above, the association with IP_{summer} of the Meuse basin is clear, which might be the reason for the lower correlation between IBF_{summer} and IP_{winter} after 1933 in Table 6.4. For the sub-period 1965–2000, IBF_{summer} of the Meuse shows a slightly higher correlation with IP_{winter} ($r = 0.67$) rather than IP_{summer} ($r = 0.57$). The downward shift in IBF_{summer} of the Semois after the early 1970s (in Figure 6.13) is striking, which is very likely related with the measurement system change (around 1967/1968) of the river. When this shift is ignored, IBF_{summer} of the Semois would to a large extent follow the evolution of IP_{summer} at Chiny.

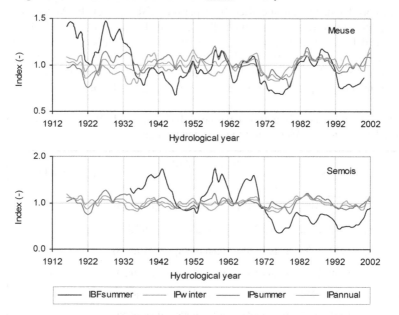

Figure 6.13 Comparison of IBF_{summer} and IP_{winter}/IP_{summer} for the Meuse (near Monsin, 1912–2000) and for the Semois (at Membre, 1930–2002), based on the 5-year moving average values (in order to minimize the effect of annual variations in precipitation on the subsurface storage). The Chiny precipitation record is used for the Semois. The unusually high value (156 m³/s) of SMIN10D in 1987 for the Meuse (see Figure 3.21) is replaced by the average value of 81 m³/s for 1986 (66 m³/s) and 1988 (97 m³/s), to remove the striking peak in the figure. Likewise for treatment of the high SMIN10D value (9.0 m³/s) in 1987 for the Semois.

Figure 6.14 illustrates the evolutions of IBF_{summer} and IP_{winter} for the Rur and the Geul. In the figure, the cause-effect relation between the two variables for the Rur seems to be affected roughly from 1978 to 1990. During this period, IBF_{summer} of the Rur was found to correlate closer with IP_{summer} ($r = 0.63$). In addition, the variations in IBF_{summer} of the Rur appear to be relatively small. The meteorological conditions in the Jeker subcatchment are expected to basically agree with those prevailing in the neighbouring Geul subcatchment. With this prerequisite, IBF_{summer} of the Jeker (1966–2001) varies mainly as a response of IP_{annual} in the area, but with a lag of one year (or longer). In Figure 3.23, the tributaries Rur, Geul and Jeker show a nearly concurrent decrease in SMIN10D over the last decade. From Figure 6.14, it appears that the period of the concurrent decrease to a large extent corresponds to a (downward) part of a natural decades-long cycle in the precipitation.

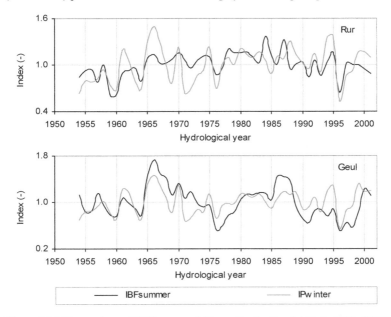

Figure 6.14 Comparison of IBF_{summer} and IP_{winter} for the Rur (at Stah, 1954–2001) and for the Geul (at Meerssen, 1954–2001), based on the annual values.

6.4.3 Relation with fluctuation of potential evapotranspiration

Similar to the definition of the precipitation index in section 6.4.2, an index for annual PET_{P-M} (denoted as $IPET_{annual}$) at De Bilt was constructed. Figure 6.15 plots the IBF_{summer} series of the Meuse and the $IPET_{annual}$ series of De Bilt for illustration. A general feature of their relation stands out: the IBF_{summer} series largely evolutes in the opposite direction of the $IPET_{annual}$ series. By combining with the relation between IBF_{summer} of the Meuse and IP_{winter} (as reflected in Figure 6.13), it was found that the periods of enhanced $IPET_{annual}$ (i.e. larger than 1), roughly from the 1930s to the 1950s and in the 1990s, mostly coincide with the periods when IBF_{summer} of the Meuse tends to be linked closer to IP_{winter}.

6.5 Hydrological simulation of the influence of forest type change

6.5.1 Introduction

In this study, the hydrological modelling study for the Meuse basin (upstream of Borgharen/Monsin) is motivated by the fact that there is a difficulty in explaining the significant decrease of the annual runoff/rainfall ratio (and some of the

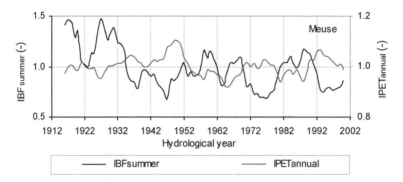

Figure 6.15 Comparison of IBF$_{summer}$ for the Meuse (near Monsin, 1912–2000) and IPET$_{annual}$ for De Bilt, based on the 5-year moving average values.

hydrological variables such as AUTD and SMIN10D) since the early 1930s, on the basis of the statistical trend results. It is expected that a detailed research into the hydrological response of the Meuse basin to climate variation and land use change with the consistent (precipitation and discharge) observation data sets could permit to draw a conclusion regarding the major cause. Recent modelling studies carried out at the basin level include De Roo *et al.* (2002) who assessed the effect of land use change with the widely physically-based LISFLOOD model, Booij (2002 and 2005) who predicted the effect of projected climate change with the conceptual HBV model, and the later HBV modelling studies, for instance, Van Deursen (2004), Arends (2005) and Leander *et al.* (2005).

The hydrological modelling of the daily discharge record for Monsin (1911–2000) was conducted by Ashagrie (2005) using the HBV model, with special reference to the influence of forest type change in the Meuse basin (upstream of Monsin). Ideally the impact of land use changes should be modelled with a physically-based model that allows for a process-based simulation of all observed changes within the basin. However, such an approach requires a strict temporal and spatial resolution that is not feasible for the study of the Meuse basin which covers a very long period of 90 years and a large basin area of 21,000 km². The HBV model is not suited to directly analyse the impact of land use changes. However, it can be used to analyse whether the observed discharge record can be reproduced equally well throughout the 20th century with available precipitation, temperature and evapotranspiration records. If this is the case, it implies that the overall impact of land use changes in the Meuse basin on the discharge at Borgharen/Monsin is small. If there exists a systematic difference between the observed and simulated discharge records, this may point in the direction of changed environments within the basin, i.e. land use changes. Besides, measurement errors, shortcomings and uncertainties of the HBV model also need to be considered. The main findings of the HBV modelling study (Ashagrie, 2005) have been summarized in Ashagrie *et al.* (2006), which constitutes the main body of section 6.5.

6.5.2 Applied HBV model and input data

Applied HBV model

The conceptual hydrological model HBV from Sweden was developed in the early 1970s (Bergström and Forsman, 1973) and has been applied in many catchments all over the world (Lindström *et al.*, 1997). The model describes the most important runoff generating processes with simple and robust routines. In the "snow routine", storage of precipitation as snow and snow melt are determined using a temperature-index method. The "soil routine" controls which part of the rainfall and melt water generates excess water and how much is stored in the soil and can evaporate. The "runoff generation routine" consists of one upper, non-linear reservoir representing fast runoff components and one lower, linear reservoir representing base flow. Runoff routing processes are simulated with a simplified Muskingum approach. HBV is a semi-distributed model and simulates the rainfall-runoff processes for each subcatchment of the basin separately.

In the applied HBV model for this study, the Meuse basin (upstream of Borgharen) was subdivided into 15 subcatchments. The model has been calibrated (1969–1984) and validated (1985–1998) on the daily basis by Booij (2002), and later fine-tuned by Van Deursen (2004) with more detailed input data from the same period. The schematisation derived from Van Deursen (2004) was used in this study. Recently the calibrated HBV model (after Van Deursen, 2004) has been used for the Meuse basin by Leander *et al.* (2005) who used generated precipitation and temperature as input for the HBV model to estimate extreme river discharges of the Meuse. It should be mentioned that the calibrated HBV model does not take into account the influence of weirs and reservoirs or other man-made alternations affecting low flows during the calibration. The simulations of low flows in the Meuse are of less quality than the simulations of normal and high flows. Arends (2005) has attempted to improve the low flow performance of the HBV model for the period 1968–1998. The author stressed that the final model should not be used to predict individual daily discharge or even multiple daily discharges (near Monsin) during low-flow periods, because the errors made are relatively large.

Discharge record (1911–2000)

The reconstructed Monsin record for the Meuse (see section 2.2.1) was used.

Meteorological records (1911–2000)

Detailed daily records for precipitation, temperature and PET are available for the Meuse basin, but cover only the short period 1968–1998 (see Leander *et al.*, 2005). The 1968–1998 detailed data set was prepared by KNMI based on the data from KMI and MeteoFrance and has only been used as reference. For the period prior to 1968, only data from a limited number of stations in or in the vicinity of the upstream basin are available for this study, including seven Belgian precipitation stations (see section 2.2.2) and five temperature stations which have long daily records (1911–2000). Four of the five temperature stations are located outside the Meuse basin. Nevertheless, it can be assumed that the temperature stations give a reasonable representation of temperature fluctuations within the Meuse basin, since temperature has a much smaller spatial and temporal variation than precipitation. Consistencies of the precipitation and temperature station records have been examined. Most stations show evidence of non-homogeneity in their station records (in terms of annual total or average), which, however, can not be explained by the available meta-data.

To avoid systematic differences between the 1968–1998 detailed data set and the 1911–2000 data set, the 1911–2000 data set was adjusted using the 1968–1998 detailed data set. Multiple linear regression equations for the common period 1968–

1998 between areal precipitation of the subcatchment (as dependant variable) and observed point precipitation at all seven stations (as independent variables) were used to generate a daily areal precipitation record (1911–2000) for each of the fifteen subcatchments. Likewise, a daily areal temperature record (1911–2000) for each of the fifteen subcatchments was generated. Finally, a new areal precipitation record (i.e. the "MeuseRLP" record in section 2.4.2) and a new areal temperature record of the Meuse basin for the long period 1911–2000 were obtained through the area weighing method. Compared with the 1968–1998 detailed data sets, for the common 31 years, the "MeuseRLP" coarse record is systematically underestimated by 5.7% and the new coarse areal temperature record is on average 2.7 °C higher.

Daily PET_{P-M} values of deciduous forest, coniferous forest, grass and cereals were calculated with the evapotranspiration module of MUST (Model for Unsaturated flow above a Shallow water Table; see De Laat and Varoonschotikul, 1996). The evapotranspiration module is based on the equation of Penman-Monteith and takes interception losses into account through including an interception reservoir in the model which fills in case of precipitation to its maximum before through-fall occurs. In this study, the available meteorological data for De Bilt (see section 2.2.3) were used in the calculations. Because of the limited data sets, a better process-based procedure for PET estimation was not possible. Based on the MUST-based PET_{P-M} values, coniferous forest potentially evaporates about 150–200 mm/a more than deciduous forest under the given meteorological conditions. PET from built-up areas was taken equal to half the value computed for grass. An areal PET_{P-M} record (1911–2000) of the basin was obtained as the area-weighted mean of the simulated values for the various types of land use. Figure 6.16 displays the annual series of two sets of areal PET_{P-M} records for the Meuse basin corresponding to both historical and current land use conditions (HLU and CLU, as defined in Table 6.5). A comparison of the MUST-based areal PET_{P-M} record with the detailed PET_{P-M} record for the common 31 years (1968–1998) shows that the averages of their annual values are very similar (546 mm/a and 559 mm/a, respectively).

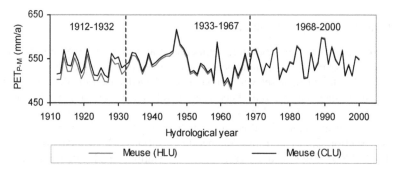

Figure 6.16 Two sets of areal PET_{P-M} records (1912–2000) calculated with the MUST model for both historical land use condition (HLU) and current land use condition (CLU) in the Meuse basin (upstream of Borgharen).

Historical and current land use/cover conditions

Detailed information on the coverages of land use categories for the Meuse basin is available only for the most recent decade (based on the CORINE data set). For the HBV modelling in this study, the historical land use (HLU) in the basin was described by the coverages of major categories for three sub-periods: 1912–1932,

1933–1967 and 1968–2000 (see Table 6.5). In the table, the coverages of deciduous and coniferous forests for 1933–1967 were estimated through linear interpolation from their historical coverages in the early 20th century and their current coverages in the current land use (CLU), while the coverages of other categories for three sub-periods were fixed, being the same as their current coverages. The historical and current coverages of deciduous and coniferous forests (with a fixed total forest area in the basin) were estimated based on the information of forest type composition in the Belgian Walloon Region given in Figure 2.14. Therefore, the HLU condition is not the real historical situation that was present in the Meuse basin.

Table 6.5 Reconstructed land use conditions in the Meuse basin (upstream of Borgharen) during the 20th century.

Coverages of land cover types	1912–1932 (previous)	1933–1967 (intermediate)	1968–2000 (current)
Deciduous forest (%)	25	21	19
Coniferous forest (%)	10	14	16
Pasture (%)[1)]	20	20	20
Arable land (%)[1)]	34	34	34
Built-up area (%)[1)]	9	9	9
Water bodies and wetlands (%)[1)]	2	2	2

Note: [1)] With the fixed coverages over the entire study period (1912-2000).

6.5.3 Results of hydrological simulations

Performance of the applied HBV model

Table 6.6 gives the summary of the model performance. For the period 1968–1998, the HBV modelling results obtained with the 1968–1998 detailed data sets can be compared with the results obtained with the 1911–2000 coarse data sets. Such a comparison will reveal to what extent the analysis is affected by a different approach to estimating the PET and the limited number of precipitation and temperature stations that are available for the entire period of interest (1911–2000). The results for the CLU condition indicate only small reductions in the values of R^2 and RE (the Nash-Sutcliffe coefficient; see Nash and Sutcliffe, 1970). The percentage bias (PB) in both cases are small. The definitions of the three criteria refer to Appendix B. From this comparison, it can be concluded that the overall performance of the HBV simulation is only slightly affected by the limited number of meteorological stations used in the estimation of areal precipitation, PET and temperature records. This conclusion is supported by the good simulation results for the entire period (1912–2000, 1911 is used to establish the initial conditions of the model). There is little difference in the model performance for the entire period when simulating the discharge under the CLU condition and the HLU condition. The simulation results for the flood volumes and the highest flood peaks (e.g. January 1926, December 1993 and January 1995) also appear good (see Figure 6.17). However, the HBV model does not correctly reproduce all observed floods, especially the medium-size floods are not always simulated well.

Table 6.6 Performance of the HBV model using different input data sets.

Period	Criteria	Detailed data set	Coarse data set – CLU condition
1968–1984	R^2 (-)	0.92	0.90
(calibration)	RE (-)	0.91	0.90
	PB (%)	5.70	5.04
1985–1998	R^2 (-)	0.94	0.93
(validation)	RE (-)	0.93	0.93
	PB (%)	2.44	1.53
1912–2000	R^2 (-)		0.91
	RE (-)		0.89
	PB (%)		2.8

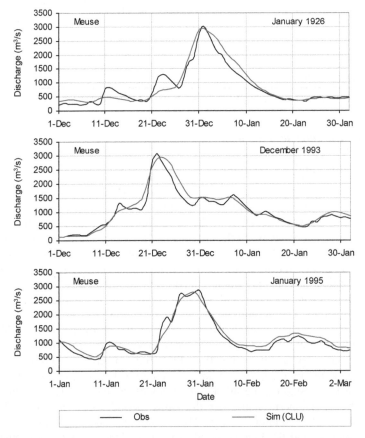

Figure 6.17 Simulated (Sim, with the CLU condition) and observed (Obs) three largest floods (January 1926, December 1993 and January 1995) of the Meuse (near Monsin) in the 1911–2000 record (source: adapted from Ashagrie *et al.*, 2006).

Sensitivity analysis

The effect of errors in the model inputs on the simulated discharge is often assessed through sensitivity analysis (e.g. McCuen, 1973). The HBV modelling study started with a sensitivity analysis of the model to known percentage changes in the input data of precipitation (by ±10%), PET (by ±10%) and temperature (by ±0.5 °C). The water balance results indicate that the applied HBV model is more sensitive to precipitation, less to PET and almost insensitive to temperature. The percentage changes in the simulated discharges due to the alterations in precipitation and PET are about ±20% and ±10%, respectively.

Change in the discharge regime

The discharge regime of the Meuse (near Monsin) is assessed in terms of annual average discharge (AAD), winter and summer average discharges (WAD and SAD), annual maximum daily discharge (AMAXD), and annual maximum 10-day moving average discharge (AMAX10D).

Figure 6.18 shows the simulated and observed 5-year moving average values for AAD, WAD and SAD in the Meuse. There appear to be systematic differences between the observed and simulated values for these variables. The simulated AAD during the sub-period prior to 1933 (without a 5-year shift) is systematically underestimated, whereas the simulated AAD during the sub-period 1933–1967/1968 is systematically overestimated. This can be observed both for WAD and SAD. The mean values of the observed and simulated average discharges (AAD, WAD and SAD) for three sub-periods are summarized in Table 6.7. Comparison of the two simulations with the HLU and CLU conditions reveals that the forest type change may only to a minor extent explain the systematic deviation in AAD, WAD and SAD for the sub-period 1912–1932, but can not explain the systematic deviation for the sub-period 1933–1967/1968.

Figure 6.19 shows the scatter plots for the observed and simulated values for AMAXD and AMAX10D. There is a tendency to underestimate AMAXD, in particular for the sub-period 1968–2000. However, as described earlier, AMAXD for the largest floods are well simulated. As shown in the figure, the observed AMAX10D values are in general well reproduced by the HBV model. Furthermore, Figure 6.20 compares the observed and simulated dates of occurrence of winter daily maxima. It shows a good simulation of the timing of most flood peaks, except for some small to moderate flood peaks. Differences of the simulation of AMAXD and AMAX10D with and without a change of forest type are negligible (not shown).

Discussion of the simulation results

The overall result of the hydrological simulation of daily discharge in the Meuse for the period 1912–2000 is satisfactory, especially when considering the fact that only a limited number of precipitation and temperature staton records were available and inconsistencies were found in the discharge, precipitation, and temperature records. However, there appear to be some systematic deviations between the observed and simulated discharge records for specific sub-periods. In the following paragraphs, the observed deviations in the HBV-simulations are discussed, with reference to the forest type change in the Meuse basin and the possible shortcomings in the available observation data.

The most obvious influence of land use on the water balance of a basin is on the

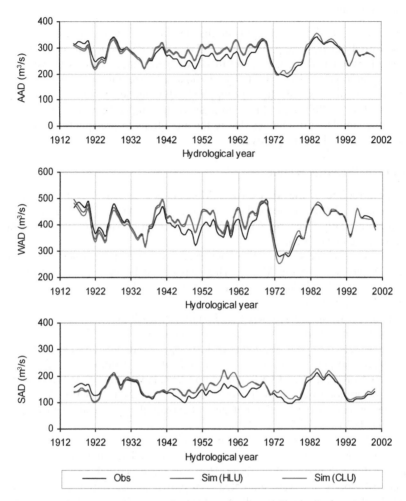

Figure 6.18 Simulated (Sim, with the HLU and CLU conditions) and observed (Obs) 5-year moving average values for AAD, SAD, and WAD in the Meuse (near Monsin) over the period 1912–2000 (source: adapted from Ashagrie et al., 2006).

Table 6.7 Mean values of the observed (Obs) and simulated (Sim) average discharges (in m^3/s) in the Meuse (near Monsin) for three different sub-periods.

Sub-period	1912–1932			1933–1967			1968–2000		
	AAD	WAD	SAD	AAD	WAD	SAD	AAD	WAD	SAD
Obs	300	432	170	265	397	135	270	400	143
Sim-HLU[1]	290	421	162	291	427	157	276	398	155
Sim-CLU[2]	282	409	157	288	422	155	275	398	155

Note: [1] Simulated with the HLU condition; [2] Simulated with the CLU condition.

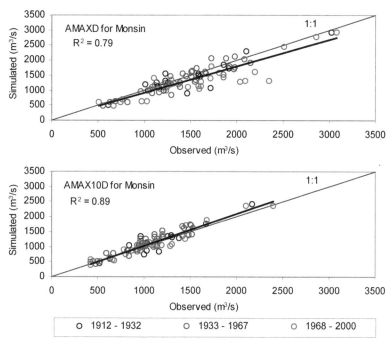

Figure 6.19 Scatter plots of the simulated (with the CLU condition) and observed floods (AMAXD and AMAX10D) in the Meuse (near Monsin) over the period 1912–2000 (source: adapted from Ashagrie *et al.*, 2006).

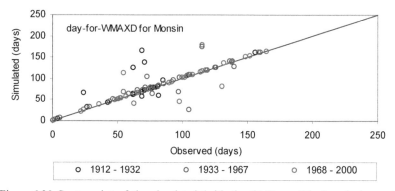

Figure 6.20 Scatter plot of the simulated (with the CLU condition) and observed day-for-WMAXD values (as defined in Figure 3.9) in the Meuse (near Monsin) over the period 1912–2000.

evapotranspiration process (Calder, 1993). However, statements about the impact of land use change on the ET volume are restricted by the accuracy of the determination of historical ET volumes. Fluctuations in the PET records may also be caused by variations in the meteorological conditions and uncertainties in the determination of ET. As reflected in Figure 6.16, the inter-decade natural fluctuation of the MUST-based areal PET_{P-M} can be up to 50 mm/a (1978–1988 vs. 1988–1998). These natural fluctuations have been taken into account in this study, but the calculation of PET introduces uncertainties in the simulation.

Table 6.7 indicates that the simulated AAD with the CLU condition during the sub-period prior to 1932 is underestimated (on average -18 mm/a), whereas the simulated AAD during the sub-period 1933–1968 is overestimated (on average +23 mm/a). Similar deviations can be observed for both WAD and SAD. The systematic underestimation of the discharge volume for 1912–1932 is reduced by about 8 mm/a in case the forest type change is taken into account. The remaining 10 mm/a (especially in the summer season) is relatively small when considering the uncertainties (expressed by standard deviations) in the precipitation and PET records used in the model study. The standard deviations of both variables for the period 1912–1932 are about 160 mm/a and 20 mm/a, respectively. Moreover, the correction factor applied to the Borgharen record for water extraction amounts to approximately 30 mm/a for this sub-period (see Figure 2.4). As noted before this fixed correction factor is uncertain especially during low flows (such as the summer of 1921) when the extraction rate was also reduced, which could not be taken into account in the Monsin record.

The deviation in the discharge volume for the sub-period 1933–1968 (see Figure 6.18) can not be explained by the forest type change using the available data sets and methods. There is no other obvious cause related to observed land use changes in the Meuse basin to explain for this systematic deviation. Another possible cause is the quality of the discharge record used. Efforts have been made to retrieve the historical meta-data of the Borgharen gauging station. The Borgharen record is based on water level measurements and a relation between water level and discharge. In order to account for changes in the geometry of the river bed (e.g. WL, 1994), the water level-discharge relation for Borgharen has been regularly updated. Around 1930 several changes took place in the Meuse near Borgharen with the construction of the Julianakanaal and the weir at Borgharen. Up to 1930, the Borgharen record is actually based on measurements at Maastricht. The impact of these changes has been accounted for in the Borgharen record. Nevertheless, based on the results presented in this analysis, one may question whether the corrections made for these impacts are always entirely correct. To further explore the cause of the deviation in the discharge volume for the sub-period 1933–1968, the Monsin record was compared with the discharge record of the Moselle at Cochem, which was provided by the German Federal Hydrological Survey (BfG). The Moselle basin neighbours the Meuse basin in the east and has comparable size and physical characteristics. As shown in Figure 6.21, the observed Moselle record corresponds very well with the simulated record for the Meuse, except for a systematic deviation for the sub-period 1933–1968. This deviation is even larger than the deviation between the observed and simulated records for the Meuse. This suggests that the deviation between the observed and simulated AAD for the Meuse may not be primarily caused by errors in the discharge record for Borgharen. A detailed analysis of the climatic input and the rainfall-runoff relation in the Moselle basin during the 20th century may help to understand the real cause for this specific period. This falls beyond the scope of this study. The above comparison also hints at a possibility of inaccuracy of the precipitation input data (i.e. the MeuseRLP record). The sensitivity analysis of the HBV model has illustrated that the simulated discharge results are very sensitive to errors in the precipitation input data. The real cause of the deviation for 1933–1968 requires further investigation. There may be other potentially unidentified land use impacts or error sources.

Figure 6.19 shows that AMAXD is often underestimated by the hydrological

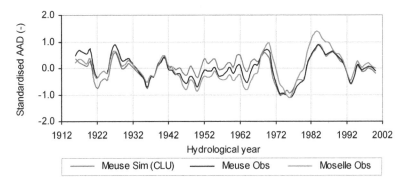

Figure 6.21 Simulated (Meuse, with the CLU condition) and observed (Meuse and Moselle)
5-year moving average values for AAD over the period 1912–2000 (source:
adapted from Ashagrie *et al.*, 2006). The mean values of AAD observed in the 91
years for the Meuse and the Moselle are 274 m³/s and 318 m³/s, respectively.

simulations. Eberle *et al.* (2002) obtained a similar result from hydrological
simulations with the HBV model for the Moselle river (1962–1997). Both Van
Deursen (2004) and Eberle *et al.* (2002) calibrated the HBV model by optimising *RE*
and *VE*. Apparently this leads to an underestimation of the medium-size floods. The
underestimation of AMAXD can be observed for all three sub-periods to some
extent, but the number of underestimated flood events for the sub-period 1968–2000
is largest. The simulation results for AMAX10D show that the total volume of the
flood is generally well reproduced by the HBV simulation. A plausible cause for the
underestimation of AMAXD may be that the flood routing module of HBV is
insufficient. Improvement requires more detailed modelling (i.e. a finer temporal
resolution, inclusion of the reservoir regulation and coupling with a hydraulic model
for flood routing) for which not all required data are readily available (Leander *et
al.*, 2005). Due to these limitations, the HBV modelling exercise performed in this
study is of insufficient detail to make statements about the impact of land use change
on the peak and shape of the flood wave. However, the simulated results
demonstrate that land use changes have not substantially influenced the volume of
floods in the Meuse during the 20th century.

6.6 Discussion

In sections 6.2 to 6.4, the rainfall-runoff relations in the Meuse basin and the
selected subcatchment have been statistically analysed with reference to different
flow states (i.e. mean flows, high flows and low flows). Furthermore, a hydrological
modelling study for the entire Meuse basin has been conducted to assist
interpretation (section 6.5). These results will be discussed for each flow state
separately. It is important to realise that the quality of the observed discharge and
precipitation data used for the Meuse basin and the selected subcatchments varies
(see sections 2.2 to 2.4). Moreover, the quality of the derived hydrological and
hydro-meteorological variables from the same observation record might vary as
well. But, it is presumed that the observation records are generally reliable for the
analysis of the rainfall-runoff relations. Certainly, the potential influence of data
inconsistencies or errors on the results can not be completely disregarded and is

considered in the interpretations. If the "observed" changes in streamflow can not be reasonably explained by climate variability or human impact, they might be related to the quality of data or limitations of the applied methodology. In the following discussion, emphasis is given to the Meuse basin.

Regarding the mean flows

The trend result of the annual runoff/rainfall ratio for the Meuse basin suggests a significant decrease (from 0.47 down to 0.41) in the runoff proportion after 1932, which, however, can not be explained by precipitation variability in the basin. In the statistical study by WL (1994), the authors also reported similar changes in the annual runoff proportion for the Meuse basin (upstream of Borgharen) during three sub-periods: 0.45 for 1911–1932, 0.39 for 1933–1973 and 0.42 for 1974–1992. An increase in precipitation amount might lead to a disproportionally larger increase in runoff due to non-linear processes. In this context, the increase (after 1974) of the runoff proportion reported by WL (1994) and the (statistically) insignificant increase (after 1978) found in the present study could be primarily explained by the increased precipitation amount (during the winter half-year) in the area (as demonstrated in Chapters 4 and 5).

Compared with precipitation, ET is less variable from year to year in the Meuse basin. Nevertheless, the decades-long fluctuation, probably attributable to variations in the meteorological conditions, is still observed, as demonstrated by the annual PET_{P-M} rate for De Bilt. It is interesting to note that both annual water loss in the Meuse basin and annual PET_{P-M} rate for De Bilt have increased coincidentally during the 1930s and the 1940s (see Figures 6.5. and 6.6). Qualitatively, there seems to be a possible link between their changes. Although the long-term fluctuation of PET_{P-M} for De Bilt should be treated with due prudence, it can provide a strong indication to the change of the actual ET in the entire Meuse basin. In the absence of a change in the annual precipitation, an increase in the PET and, consequently, in the actual ET, leads to a decrease in the runoff proportion. In this context, the decrease of annual runoff/rainfall ratio for the Meuse basin around 1932 might be at least partly attributable to the enhanced PET rate. Moreover, the dominant forest type in the Meuse basin has actually changed during the 20th century from deciduous to coniferous trees (see section 2.7.2). Under certain climatic conditions, coniferous trees generally have the highest transpiration and interception losses. However, The absence of an obvious change in the quarter-flow interval of the Meuse during the 20th century suggests that any change in the soil moisture storage after the dry period in the basin (upstream of Monsin), due to the effects of the enhanced PET rate and the forest type change, is too small to be (statistically) detectable. Furthermore, the HBV-simulations using the consistent (precipitation and discharge) observation data sets also show that there appear to be some systematic deviations between the observed and simulated discharge records for the specific sub-period 1933-1967, for which the real cause is not yet clear. A more detailed process-based analysis is required to quantify the actual impacts of forest type change.

Statistical results of annual runoff/rainfall ratios for the selected subcatchments indicate no systematic changes in their rainfall-runoff relations. Evaluation of the rainfall-runoff relations in the Jeker subcatchment is hampered by lack of adequate precipitation data in this study. Given small difference in the precipitation conditions in the region, the variations in the annual runoff of the Jeker subcatchment (1965–2001) have been found to largely reflect the variations in the annual precipitation (based on the Geul precipitation record) with a lag of one year.

Regarding the high flows

The concurrent increase of the (winter) peak discharges in the Meuse and the selected tributaries (excluding the Geul and the Jeker) over the last couple of decades (see section 3.3.2) suggests a climatic link. This has been well demonstrated by the effect of antecedent 10-day precipitation depth (A10P-WM/WPOT) on the magnitude of (winter) floods in the Meuse. Meanwhile, there seems little change in the (winter) flood volumes in the Meuse, as suggested by the analysis results of winter k-day maximum runoff in the river (see section 3.3.3). Furthermore, the HBV-simulation results confirm that the recent increase of the (winter) peak discharges in the Meuse could be largely ascribed to climatic variability. The hydrological consequence of increased (winter) precipitation amount and intensity since 1980 could be more pronounced in the hilly Semois subcatchment than in the large Meuse river basin. However, due to the possible influence of the data inconsistency, the absolute percentage change in the flood magnitude (e.g. a striking increase by 67% for AMAXD after 1979, see Figure 3.8) of the Semois appears to be much higher than the expected change, and therefore should be cautiously taken.

The climatic cause of the increased peak discharges in the Meuse (and the selected tributaries excluding the Geul and the Jeker) over the recent two decades agrees with or is partly supported by the findings of a number of flood trend studies in the Meuse basin and various European countries. WL (1994) analysed the 83 highest peaks (on average 1665 m^3/s) in the Meuse (at Borgharen) in the period 1911–1993. Higher peaks were found in the latter 21 years. Double mass analysis of the peaks above 1500 m^3/s and the 10-day precipitation totals before the peaks indicated that after 1960 more precipitation appeared to come to runoff, leading to an increase in the peak discharge by 17%. However, no increase was shown in the peaks between 1000 and 1500 m^3/s. Pfister *et al.* (2000) reported a significant increase in the winter daily maxima of the Alzette river at Esch (1954–1996) since the 1980s, which could be related to a marked increase in the winter precipitation since the 1970s. Uhlenbrook *et al.* (2002) demonstrated the precipitation as the most significant factor (compared to other basin properties) for causing floods (return periods larger than ten years) in 29 meso-scale catchments (76–4,044 km^2) in Southwest Germany. Caspary (1995) found that the non-homogeneity in the annual peak discharges of the Enz river (1932–1995) and the upper Danube river (1926–1995) around 1976 was caused by a changed winter climate. In UK, Robson *et al.* (1998) examined trends in the national flood behaviour (flood size and frequency of occurrence) using extensive POT and annual maxima data over the period 1941–1980/1990 and found no regional flood trends. Analysis of a limited number of stations over the past 80–120 years also indicated no statistical evidence of a long-term trend in flood peaks. The inter-annual variations in UK floods appeared to be climatically driven. EEA (2001) summarized that in the medium and large-size river basins in northern and central Europe, wide-ranging and continuous precipitation was commonly the main factor in flood generation. Pfister *et al.* (2004) reviewed that due to the increased winter precipitation totals and intensities over the second half of the 20th century, signs of increased flooding probability in many areas of the Rhine and Meuse basins have been documented.

Regarding the low flows

Graphical analyses of precipitation, PET and summer low flow (SMIN10D) for the Meuse basin and the selected subcatchments show that most variations in SMIN10D are linked to variations in the meteorological conditions (but with different lags in time), except for the distinct downward shifts in SMIN10D of the Meuse (around 1933) and in SMIN10D of the Semois (around 1971). The lag in time depends very much on the storage capacity of the catchment. For example, the dominating influence of current (i.e. summer half-year) meteorological conditions on SMIN10D in the Semois is attributable to low storage capacity of the subcatchment. In contrast, the lag of one year in the reaction of SMIN10D in the Jeker is associated with the large groundwater storage in the chalky aquifer.

An increase in PET could reduce groundwater recharge in the Meuse basin during the winter season and, consequently, affect the magnitude of low flows during the subsequent summer season. On the basis of the enhanced PET, the winter recharge in the area would have been reduced during the specified periods roughly from the 1930s to the 1950s and in the 1990s. However, the actual impact on the summer low flows of the river is also affected by the summer precipitation conditions. It is worth noting that the early period prior to 1933 is not special during the 20th century, regarding the precipitation and PET conditions. This implies that the marked decrease in SMIN10D of the Meuse around 1933 might be primarily caused by other factors rather than climate-linked variations or fluctuations in precipitation and PET. The likely explanations are inaccuracy of the Monsin record for the earlier years prior to 1933, which has been mentioned in section 3.6, and human impact, for example, resulting from forest type change in the area. The marked shift in SMIN10D of the Semois around 1971 appears to be largely caused by the measurement system change (around 1967/1968) of the river. This effect appears to overshadow the influence of the drier summer climate since the late 1960s. In the climate scenarios impact study, Gellens and Roulin (1998) demonstrated that the Semois subcatchment is very sensitive to low flow situation and is likely to experience a more severe low flow situation (more low flow days) in future due to its low storage capacity and its fast response. The short-term decreases identified for SMIN10D in the Geul and the Jeker in the last decade (see section 3.5.2) seem largely attributable to the decades-long cycle in (winter) precipitation and should not be viewed as real alterations. It is not sure whether the above interpretation also applies to the concurrent decrease of SMIN10D in the Rur, because the river has been subject to regulation of the reservoirs throughout the whole period of record and also the impact of groundwater abstractions in the area, which is, however, difficult to assess.

Low flows and droughts are regional in nature, as a consequence of hydro-climatological influences (Demuth, 2005). The research results of the EU supported project ARIDE (Assessment of the Regional Impact of Droughts in Europe) have led to a series of publications on the low flow and drought topics at national and European levels (e.g. Demuth and Stahl, 2001). In spite of the fact that many European river basins have experienced severe droughts during the recent three decades (e.g. 1973, 1976, 1988–1992 and most recently 2003), the trend analysis performed by Hisdal *et al.* (2001) reveals that there is no clear indication that streamflow drought conditions in Europe have generally become more severe or frequent in the second half of the 20th century. Enhanced ET may act in combination with decreased precipitation to aggravate the severity and duration of the drought event (EEA, 2001). According to Demarée *et al.* (2002), during the 20th century there was evidence of an increased temperature in Central Belgian which started from the year 1930 onwards and ended with the well-known warm 1990s. Gellens and Schädler (1997) reported a possible rise in the ET rate equivalent to some seven to ten percent in a set of eight catchments in Belgium due to projected climate change. Brouyère *et al.* (2004) conducted a study of climate change scenarios impact on the chalky aquifer of the Jeker subcatchment and predicted a decrease in groundwater levels and reserves on a multi-annual basis. The above-cited statements and findings in the literature help to understand and to substantiate the results of summer low flows in the Meuse and the selected tributaries.

6.7 Concluding remarks

Based on the statistical results and the discussion above, the following concluding remarks can be drawn:

- During the 20th century, the annual rainfall-runoff relation in the Meuse basin (upstream of Monsin), described by the annual runoff/rainfall ratio, appears to have changed around 1932 (with a significant decrease in the runoff proportion). The annual relations in the selected subcatchments remain relatively stable during their study periods.

- Potential evapotranspiration (PET) in the Meuse basin, described by PET$_{P-M}$ for De Bilt, shows climate-induced fluctuations. The enhanced PET (mainly March to August) occurred roughly in the 1930s to the 1940s and the 1990s.

- The relatively large magnitudes of (winter) floods in the Meuse and the selected tributaries (except for the Geul which shows an apparent decrease in the flood magnitude since 1971) in the last couple of decades are largely affected by the increased antecedent precipitation depths, and thus can broadly be ascribed to climate variability.

- Fluctuations of summer low flows (SMIN10D) in the Meuse and the selected tributaries are in general in response to variations in the meteorological conditions (with different lags in time), except for the distinct shifts shown in the Meuse (around 1933) and the Semois (around 1971). The possibility of human impact (e.g. reservoirs, groundwater abstractions) on SMIN10D of the Rur can not be discarded.

Besides climate variability, human intervention (particularly land use change in the upstream Meuse basin) is another possible (and often criticized) cause for alteration of the rainfall-runoff relation. Evaluation of the potential human impacts on the runoff behaviours of the Meuse river and the selected tributaries is summarized in Chapter 7.

7   Effects of human activities on the discharge regimes

7.1 Introduction

Land use activities may affect streamflow through various hydrological processes and pathways. In section 2.7, historical land use changes and hydraulic activities in the Meuse basin and the selected subcatchments during the 20th century have been described. This chapter intends to make assessment of their effects on the discharge regimes of the rivers (emphasizing the Meuse river basin). Seven sections are organised. In sections 7.2 to 7.4, the potential impacts on mean flows, high flows and low flows of the rivers are separately evaluated in a qualitative way based on both the statistical trend results and the HBV-simulated results. The existing knowledge of human impacts on streamflows (see previous section 1.2.2) and the results of recent relevant studies mainly restricting to the Meuse basin and the European areas (see sections 7.2.1, 7.3.1 and 7.4.1) provide a sound scientific basis for the evaluation. Section 7.5 makes a synthesized evaluation of the hydrological effects of human activities in the Meuse basin. Finally, section 7.6 presents the concluding remarks and also the suggestion of further researches.

7.2 Effect on the mean flows

7.2.1 Introduction

Studies of human impacts on annual rainfall-runoff relations (or annual mean flows) in the Meuse basin and the subcatchments thereof are limited. Based on the statistical results in WL (1994), the authors have briefly discussed the potential human impacts on the river flows of the Meuse (upstream of Borgharen). They suggested that the increased runoff proportion after 1974 was possibly a reflection of rapid urbanisation, but they found it difficult to explain the higher runoff proportion in the earlier years prior to 1933 due to historical forest area change in the basin. Bultot *et al.* (1990) conducted a hydrological modelling exercise for the Houille subcatchment (114 km^2) in the Ardennes. The simulations of different land use scenarios with the IRMB model show that for the study period 1901–1984, an increase of coniferous forest area would result in a decrease of annual flow. The largest difference in impacts on the annual flows was found between the basin-wide coverages by coniferous forests (556 mm/a) and pastures (631 mm/a) as a result of the maximum ET rate (552 mm/a) and the minimum ET rate (477 mm/a). Intermediate results were obtained for other vegetation types (e.g. deciduous forests and crops). Similar studies for other European river basins or areas, either based on the observed data or based on the simulated results, are considerable in literature. Seuna (1999) reported the hydrological effects of forestry treatments (namely forest harvesting and draining) in Finland. As a result of draining, total runoff increased during the first ten years, but after 15 to 20 years the total runoff returned to its original level.

In the following sub-sections, the potential impacts of human activities on annual runoff volumes (annual mean flows) and runoff regimes (flow regimes) of the selected subcatchments and the Meuse basin are assessed.

7.2.2 Selected tributaries

Semois river

Different from the general trend of increased coniferous forest area in the Ardennes, a large part of the forestland in the Semois subcatchment is still occupied by deciduous forest. According to the available (limited) information, there seems no significant land use change occurring in the subcatchment during the study period. The reduction of agricultural area over the recent two decades (likely converted to the urban area) is minor. The overall impact of land use changes in the area on the annual runoff appears to be marginal or insignificant. The assessment is seemingly supported by the statistical results of annual mean flow (AAD) and annual runoff/rainfall ratio for the Semois (see Figures 3.2 and 6.3), which indicate no significant change throughout the period of observation. Nevertheless, one may also argue that the hydrological response of the hilly Semois subcatchment to the changing climate (with enhanced annual precipitation) in the last two decades should be more pronounced than the basin-average level. The absence of an (obvious) increase in the annual runoff proportion might be caused by the shortcoming of involving the single Chiny station record (not an areal precipitation record) in calculation of the runoff/rainfall ratio series.

It is possible that the annual runoff volume remains stable, but the runoff regime has altered due to the changing precipitation regime within the year, the impact of human activities, or simply the measurement system change. Here the monthly runoff distributions for the different sub-periods are compared (see Figure 7.1). The year 1980 in Figure 7.1 is generally viewed as the timing of precipitation change in the entire Meuse basin (as demonstrated in Chapters 4 and 5), while the other year 1968 corresponds to the timing of the measurement system change for Membre on the Semois. It is worth mentioning that the climate during the 1970s was abnormally dry, thus influencing the magnitude of runoff regime as a whole. As reflected in Figure 7.5, the effect of precipitation change on the seasonal runoff behaviour of the Semois is observable, e.g. increased winter (DJF) runoff as a result of enhanced winter precipitation after 1980 and increased runoff in March mainly attributable to increased precipitation in that month over the 1980s. The decreased runoff in August and September after 1968 is not only a reflection of drier summer starting from the late 1960s, but also likely subject to the influence of the measurement system change of the river. Recall that SMIN10D in the Semois shows a distinct downward shift (around a similar year) which could not be explained by the meteorological factors (precipitation and ET). The Vierre reservoir (constructed in 1967, see Figure 2.22) is too small to affect the seasonal flows of the Semois.

Rur river

During the second half of the 20th century, the large-scale lignite mining and the construction of large reservoirs constitute the most important human interventions in the hydrological system of the Rur subcatchment. The operation of the reservoirs generally has little effect on the annual runoff. Therefore, the evaluation

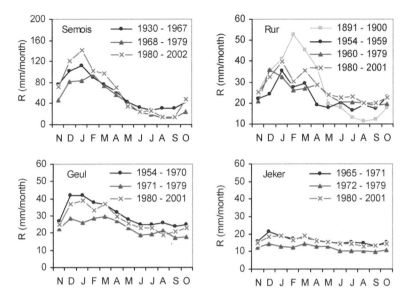

Figure 7.1 Changes in the runoff (R) regimes of the selected tributaries Semois (at Membre), Rur (at Stah, only 1891–1900 for Vlodrop), Geul (at Meerssen) and Jeker (at Nekum).

concentrates on the potential impact of the mining activities. The annual rainfall-runoff relation in the Rur subcatchment has been found to be stable over the study period since 1954 (see Figure 6.4). However, one may wonder whether the subcatchment is subject to a certain amount of water loss throughout the period of observation, because a large amount of groundwater was continuously extracted for the mining activities (although most of the extracted groundwater returned into the downstream river). Moreover, according to the information collected from WVER, groundwater in the deep aquifer of the Rur subcatchment can naturally flow eastward into the neighbouring Erft catchment as a result of the difference in hydraulic heads. However, it is difficult to quantify the amount of groundwater crossing the fault system into the Erft catchment.

To investigate the possible water loss in the Rur subcatchment, the record of daily water release from the dam site at Heimbach (see Figure 2.21) for the study period (1953–2001) was collected from WVER. Presume that the influence from other small tributaries between Heimbach and Stah on streamflow downstream is small. A comparison of the annual mean discharge (AAD) at Stah and the annual average water release (AAD) at Heimbach would possibly expose the potential impact of the mining activities. Figure 7.2 displays the AAD series for two stations, together with their ratio series. In the figure, variations of two AAD series look very similar except for the discrepancy being slightly closer in the early years prior to 1960. Statistical analysis of the entire ratio series by the Pettitt test indicates that the change point year 1959 (after pre-whitening) is significant with a high probability of 0.91. Further analysis of the ratio series for the remaining period 1960–2001 shows a possible change around 1989 with a low probability of 0.67 (not statistically significant). There are several possible explanations of the relatively low AAD values (underestimated by several m^3/s) for Stah from 1954 to 1959. It could be seen as an indication of a possible mining-related water loss in the Rur subcatchment, whereas from 1960 onwards a new equilibrium of the hydrological system seems to have established. Another possible explanation is related to the potential inconsistency of the Stah record arising from composition of two short records (1953–1960 for Drie Bogen and 1960–2001 for Stah, see section 2.3.4) around 1960. In addition, the quality of the Heimbach record may be problematic. Detailed information of this discharge record is not available for this study.

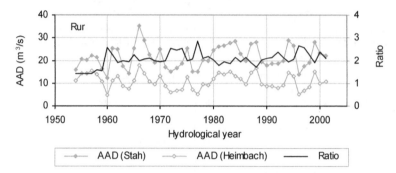

Figure7.2   Time series of AAD at Stah and AAD at Heimbach on the Rur, together with time
series of their ratio (1953–2001).

In Table 2.1, it has been noted that a short record of daily discharge for the last decade of the 19th century (1891–1900), together with an incomplete daily record from 1953 to 1971, was available for the Rur at Vlodrop (some 5 km downstream of Stah). Contrasting the recent record with the historical record would reveal whether and how far the "natural" discharge of the Rur has been altered as the consequence of the human activities occurring in the 20th century (e.g. reservoirs, lignite mining and land use changes etc.). Figure 7.3 plots the AAD series for Vlodrop over the two different periods (1892–1900 and 1954–1965) together with the AAD series for Stah over the latter period (1954–1965). There appears to be an overall consistency between the Stah record and the Vlodrop record over the common period. This means that AAD at Vlodrop in the historical period can be considered as an approximation of AAD at Stah in that period. From the figure, AAD of the Rur seems nearly stable throughout the historical and latter periods. The mean values of AAD for Vlodrop are almost the same, around 23 m^3/s. The available precipitation record for Aachen (near the western boundary of the Rur subcatchment) fortunately covers the historical period, thus allowing any observed hydrological situation in the Rur to be interpreted in the context of climate variability. Over the common period 1954–2001, annual precipitation total (APT) at Aachen (on average 820 mm/a) has been found to be in general consistent with APT in the Rur subcatchment (on average 836 mm/a). Another historical station Roermond (at the mouth of the Rur) gives a different mean value of APT (on average 720 mm/a) and therefore was not used. The APT series for Aachen is added to Figure 7.3. It shows that the annual precipitation conditions in the two periods of concern are similar, the mean values of APT (at Aachen) being 846 mm/a and 832 mm/a, respectively. The above step change analysis results imply that there is no detectable change in the annual rainfall-runoff relation in the Rur subcatchment during the 20th century. The hydrological effect of human activities in the area is not evident. The relatively low values of AAD at Stah prior to 1960 can not be convincingly seen as evidence of the hydrological effect of the mining activities, especially when considering the very small amount of underestimation and uncertainty in the Heimbach record. To gain further insights into the impacts of human activities (with particular attention to reservoir construction, land use changes and lignite mining) in the Rur subcatchment, Yusuf (2005) has conducted a HBV-modelling study of daily discharges in the last decade of the 19th century (1891–1900) and the recent decade (1992–2001) using low-quality input data (e.g. areal precipitation estimated from only two stations Aachen and Roermond). The simulated results of scenarios suggest that the decrease of water losses, which could be largely attributed to open pit mining activities, is very small (-12 mm/a).

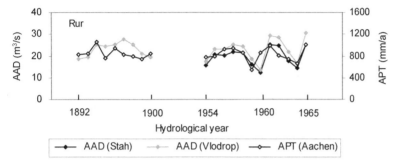

Figure 7.3 Time series of AAD in the Rur (at Stah and Vlodrop) and APT at Aachen over two different periods (1892–1900 and 1954–1965).

The natural flow regime of the Rur over the second half of the 20th century has obviously altered due to regulation of the reservoirs in the upstream area (see Figure 7.1). In the graph for the Rur, the year 1960 corresponds to the enlargement of the largest reservoir Rurtalsperre (see Figure 2.21) and also coincides with the potential inconsistency of the Stah record. The flow regime in the historical period (1891–1900) reflects the natural flow regime before large-scale human interventions. After construction of the reservoirs starting from the early 20th century, the monthly mean flows from February to April have decreased, whereas the monthly mean flows from July to September have obviously increased. The hydrological effect of wetter winter climate over the last two decades on winter seasonal flows is likely to be partly weakened by regulation of the reservoirs within the year.

Geul and Jeker rivers

The main changes in land use and water management in the Geul subcatchment over the second half of the 20th century are intensification of agricultural land (e.g. causing an obvious change in crop types), urbanisation (e.g. increasing the urban area by 5%) and groundwater abstraction. Detailed information indicates that the sewage water from some towns together with the flood water stored in three retention areas are routed to some place outside the subcatchment (see Agor, 2003). In view of no apparent change in the annual runoff/rainfall ratio series (see Figure 6.4), the combined effect of the forementioned human activities on annual runoff from the Geul subcatchment may be too small to be detected.

There is a clear indication that the Jeker subcatchment is subject to a great amount of water loss due to groundwater lateral outflow and intensive groundwater abstraction throughout the entire period of record. For the period 1966–2001, the mean value of annual area specific runoff (mm) from the Jeker subcatchment accounts for only 50 to 60% of the mean values for the neighbouring Rur and Geul subcatchments.

Likewise, Figure 7.1 presents the monthly runoff distributions for the two subcatchments. In the graphs concerned, the year 1971 for the Geul and the year 1972 for the Jeker correspond to their measurement system changes. Probably affected by the data inconsistency, the monthly runoff values for the Geul after 1971 show an overall tendency of decreasing to some extent, which makes the hydrological effect of wetter climate in winter and early spring after 1980 not clear. Due to the strong regulation capacity of the groundwater system, the flow regime of the Jeker is damped and less sensitive to seasonal precipitation change, land use change and local groundwater abstraction in the catchment area.

7.2.3 Meuse river

The hydraulic activities on the river channel of the Meuse might have had pronounced influences on flood flows or low flows in the river, but they generally should have little effect on annual mean flows of the river. Influence of water extraction by canals between Liège and Borgharen has been taken into account in the Monsin record. In this sub-section, attention is given to the potential impact of land use changes in the Meuse basin (upstream of Monsin) on the annual runoff (the annual mean flow).

During the 20th century, the most important land use changes in the Meuse basin, which also potentially have most significant effects on the annual runoff volume, are the forest type change in the Belgian Ardennes since the late 1920s and the expansion of urban areas since the 1950s. The forest type change with an increased proportion of coniferous forests tends to reduce annual runoff from the basin mainly due to increased ET loss, whereas the urbanisation tends to increase annual runoff from the basin due to increased impervious area. However, the HBV-simulated results (section 6.5) clearly demonstrates that the overall impact of land use changes on the annual runoff from the basin appears to be marginal during the 20th century. The hydrological effects of management practices associated with forestry and agriculture (e.g. roads, drainage, ditches in the field) in the Meuse basin are difficult to evaluate, but they seem unlikely to be linked to the observed changes in the runoff proportion. Considering that the extent of the urban area (currently 9%) is still limited relative to the entire basin area (upstream of Monsin), the resulting effect on the annual runoff volume would be least significant in such a large basin. Even in the small Geul subcatchment, the effect of urbanisation did not show up over the study period, possibly due to its small percentage (from 6% up to 11%) in the area. Therefore, the slightly higher runoff proportion in the Meuse basin after 1974 (or 1978) is more likely associated with the wetter winter climate. This interpretation disagrees with the explanation (related to urbanisation) given by WL (1994).

Due to the relatively small storage capacity of soils and the relatively steep slope in the large part of the Ardennes, the flow regime of the Meuse is highly a reflection of the precipitation regime. The effect of the reservoirs and land use changes in the upstream Meuse basin on the flow regime is hardly observable. Figure 7.4 compares the monthly runoff distributions of three different sub-periods. As mentioned earlier, the year 1980 is viewed as the timing of precipitation change in the entire Meuse basin. Besides, this year also roughly coincides with the occurrence of a recent insignificant increase (around 1978) in the annual runoff/rainfall ratio series of the Meuse (see Figure 6.3). Another year 1932 corresponds to the occurrence of a major decrease in the annual runoff/rainfall ratio series. Due to the enhanced precipitation amount in the last two decades, the runoff volumes in the winter season (mainly December and January) and in the early spring (March) appear to have obviously increased, relative to the middle sub-period (1932–1979). However, there is a similar problem of explaining the higher runoff values of November to January in the early sub-period (1912–1931), given little difference in the precipitation conditions of these months before and after 1932 (98 mm/month vs. 91 mm/month). The decrease in the monthly runoff of August and September due to drier summer from the late 1960s is still observable at the basin level, as reflected in the flow regime of the recent sub-period (1980–2000).

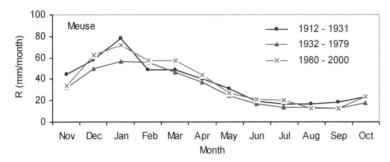

Figure 7.4   Changes in the runoff (R) regime of the Meuse (upstream of Monsin) during the different sub-periods (1912–1931, 1932–1979 and 1980–2000).

7.3 Effect on the high flows

7.3.1 Introduction

Recent studies of land use effects on floods in the Meuse basin were largely stimulated by the severe flood events in the 1990s (mainly December 1993 and January 1995) and the need of improved flood risk management. In the case study of the Houille subcatchment in the Ardennes, Bultot *et al.* (1990) has shown that the simulated flood peak discharge was almost unaffected by land cover change. In another two Ardennes subcatchments Ourthe Orientale (317 km^2) and Mehaigne (343 km^2), De Wit *et al.* (2001) found that a complete afforestation with deciduous forests might reduce peak flows of the rivers. In the Geul subcatchment, Leenaers and Schouten (1989; also see Dautrebande *et al.*, 2000) claimed that urbanisation and modernisation in agricultural practice in the area during the last 50 years have contributed to higher peak discharges in the river. Through modelling, Dautrebande *et al.* (2000) demonstrated that the proposed land use adjustment measures could only weakly reduce the (hourly) peak discharge in the Geul, maximally by 10% compared with the historical situation in the 1950s. Based on the sedimentation rates of floodplain deposits in the Geul subcatchment, Stam (2002) found that mining activities and deforestation in the 19th century caused more small floods. There are several studies specifically dealing with the basin-scale effects (e.g. WL, 1994; De Roo *et al.*, 2002). In the interpretation of the statistical results, WL (1994) doubted whether the increasing impervious area in the Meuse basin was the reason for the increase (by 17% after correction for differences in rainfall) of peak discharges above 1500 m^3/s, because the increase of impervious area generally affects in particular the smaller and more frequent peaks. De Roo *et al.* (2002) applied the LISFLOOD model to simulate the 1995 flood event in the Meuse (at Borgharen) with the land use patterns in 1975 and 1992 (with a slight increase in the urban area and a slight decrease in the forested area). The simulations suggested that the peak discharge and the flood volume simulated for 1992 were both higher by 0.2 % and 4.06 % as compared to the values for 1975. The authors also mentioned that the results should be interpreted with care, given the uncertainty in the model input data. In a recent short paper, De Wit *et al.* (2003) summarized the most recent results of model calculations (using the LISFLOOD model) in the Meuse basin. The conclusions are that due to the effect of dissipation by channel routing and human intervention such as weirs, canals etc. in the river, the discharge of the downstream Meuse river may not be altered by effect of land use change so much; moreover, the proposed flood meanagement measures in the French and Belgian parts (including reforestation, re-meandering, increased inundation of floodplains, nature recovery in small streams and uncontrolled retention) only have a limited effect on the peak discharge at Borgharen during extreme events.

In recent two decades, a number of land use impact studies have been conducted in the Rhine river basin and other European river basins with a particular concern of flooding. Some examples are described here. In the Steinsel subcatchment (408 km^2) of the Alzette basin (Luxembourg), Liu *et al.* (2004) and Gebremeskel *et al.* (2005) applied the GIS-based distributed hydrological model WetSpa to assess the runoff contributions from different land use classes and the potential impact of land use changes on runoff generation. Simulation results show that afforestation has a mild positive effect in reducing the (hourly) peak discharge in comparison to the present situation (by 5.3%), whereas urbanization and deforestation lead to an increase of the simulated (hourly) peak discharges (by 26% or a lesser extent). The magnitudes of changes, however, differed from one storm to another depending upon the antecedent soil moisture content. In a modelling study of several meso-scale catchments (hundreds km^2) in Germany, Niehoff *et al.* (2002) have shown the influence of the land use on storm runoff generation for different types of rainfall events. The influence was most distinct for short, intense rainfall events and minor for longer, less intense rainfall events. A simple alteration in the interception capacity did not yield significant changes in catchment response; more pronounced changes arose from modifications in the infiltration conditions, particularly sealing of the soil surface. Similar findings were also reported by Ott and Uhlenbrook (2004) for a catchment in Southwest Germany. Pfister *et al.* (2004) reviewed the recent modelling results in the Meuse and Rhine river basins. A very limited effect of land use changes on the discharge regime seems to exist for the main branches of the Meuse and Rhine rivers. Land use changes, particularly urbanization, can have significant local effects in small headwater subcatchments, especially during intense local rainstorms, but no evidence exists that land use change has had significant effects on peak flows in the Meuse and Rhine rivers. In Norway, Tollan (2002) reported that changes in forestry and the state of forests did not lead to measurable changes in the floods of the main Glomma river (with a drainage area of about 42,000 km^2) over the period 1920–1990. The simulated results of case studies in small subcatchments within the river basin have shown that a complete deforestation could increase flood peaks by nearly 60%; urbanisation would cause an average increase in the flood volume by the order of 20% for the 1995 flood event (compared to the pre-urban state) and an increase in the frequency of floods. Impacts of land management practices and hydraulic engineering activities on flooding also receive concern in the literature. Bronstert *et al.* (1995) reviewed the impact of land consolidation on runoff production and flooding in Germany. The conclusion is that land consolidation may influence local runoff production processes to a significant degree, whereas for large river catchments, land consolidation has no important effects in flood characteristics. Lammersen (2000) found that river training and retention measures along the German Rhine could have remarkable impacts on the peak discharges in the Lower Rhine region (in Germany and The Netherlands), significantly depending on the pre-flood conditions in the catchment and the temporal and spatial distribution of the rainfall in the Rhine basin. Tollan (2002) reported that the increase in peak discharge caused by the flood protection works along the Glomma river amounted to about 1.5% in the lower end of the protected reach. Mudelsee *et al.* (2003) concluded that reductions in river length, construction of reservoirs and deforestation in the past centuries have had minor effects on flood frequency of the Elbe and Oder rivers in central Europe.

In the following sub-sections, the potential impacts of human activities on winter flood behaviours of the selected tributaries and the Meuse are assessed. Summer floods in the Meuse basin often occur locally and hence are less important to the flooding risk of the downstream Meuse river. In general terms, summer flood behaviours of the rivers are more sensitive to land use changes, because they are affected more by intense rainstorms which can generate local overland flow.

7.3.2 Selected tributaries

Semois river

The magnitude of winter floods (e.g. WMAXD and WPOT) in the Semois has been found to show a remarkable increase in the last two decades, along with more frequent flood events. Over the period of increased flood severity and frequency in the Semois, a slight reduction of agricultural area in the subcatchment, which is likely to be a hint of urbanisation, has also taken place. However, considering that the extent of urban area in the catchment area is still very low and the extreme floods (e.g. $WPOT_{158}$) often occur in the wet mid- and late-winter season (i.e. January and February), effect of the urbanisation in the last two decades would be weak or negligible. Climate variation is the dominating factor for the changed flood regime of the Semois since 1979. The absolute percentage change for instance in WMAXD of the river (about 68%) appears to be relatively high, possibly to some extent subject to the influence of the measurement system change (for Membre) around 1967/1968. The computed percentage change of flood peak discharge (and frequency of flood events) for the Semois should be used with great caution. The Vierre dam as well as its upstream area is too small to impact the flood regime of the Semois. Both the day-for-WMAXD series and the quarter-flow interval series for the Semois show a relatively stable feature over the period of observation, suggesting no obvious change in the water-retention capacity in the area.

To illustrate the change of high flows in the Semois in detail, Figure 7.5 plots the flow duration curves of winter half-year daily discharge in the river (at Membre) for different sub-periods (with an equal length). It clearly shows that the magnitude of winter daily flows with a duration less than 30% (accordingly larger than 50 m^3/s) has obviously increased in the recent sub-period 1979–1999. The change in the high flow condition before 1970 is marginal. Using the Chiny station record, the mean values of winter half-year precipitation total (WPT) computed for 1930–1949 and 1950–1970 are 683 mm/period and 724 mm/period, respectively. The slight difference in the winter precipitation condition seems a reasonable explanation of the higher daily flows with a duration less than 10% occurring in the sub-period 1950–1970.

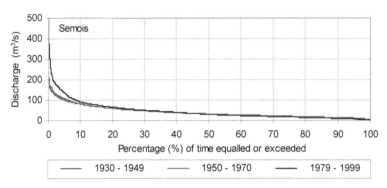

Figure 7.5 Comparison of flow duration curves of winter half-year daily discharge in the Semois (at Membre) for three different sub-periods.

Rur river

The magnitude of winter floods in the Rur is subject to the influence of the large reservoirs in the upstream area. Figure 7.6 plots the WMAXD series of the river before and after construction of the dams to illustrate the attenuation. As reflected in the figure, there is an apparent reduction in WMAXD of the Rur at Stah (Vlodrop) since the 1950s, as compared with WMAXD at Vlodrop in the historical period (1892–1900). The mean value of WMAXD at Stah (Vlodrop) for 1954–1965 is 63 m³/s (71 m³/s), being less than a half of the historical mean level of 161 m³/s at Vlodrop. Given uncertainty (e.g. measurement error, instrument change etc.) in the historical observation data for Vlodrop, the above absolute discharge values are preferably interpreted in a qualitative way. To examine whether there is also a change in the flood runoff volume, an additional analysis of WMAX60D in the Rur has been done. After smoothing of the flood peaks, no significant step change of WMAX60D between the early and late periods was found (see Figure 7.6). Computation of WPT at Aachen for the two periods (on average 846 mm/period for 1891–1900, and 832 mm/period for 1954–1965) indicates only a small difference in WPT over time (see Figure 7.6). This means that the distinct step change observed in WMAXD of the Rur is not merely a reflection of change in the winter precipitation condition. The HBV-modelling study by Yusuf (2005) has demonstrated the attenuating effect of the reservoirs on flood peaks in the Rur.

The above analysis results prove that the magnitude of winter floods in the Rur recorded in the second half of the 20th century has been considerably reduced due to construction of the dams which started from the early 20th century. However, the regulation of the large reservoirs does not seem to have significantly affected the climate-induced increase of winter peak discharges in the Rur over the recent two decades (see Figure 3.8). In addition, the effect of a substantial increase in the total capacity of the large reservoirs after 1960 (due to enlarging the Rurtalsperre reservoir) is not detectable from the data used in this study. Due to the large regulation capacity of the reservoirs, effect of land use changes in the Rur subcatchment on the flood regime of the river, if any, would be least significant.

Geul and Jeker rivers

In the Geul subcatchment, the substantial decrease of winter flood peak discharge

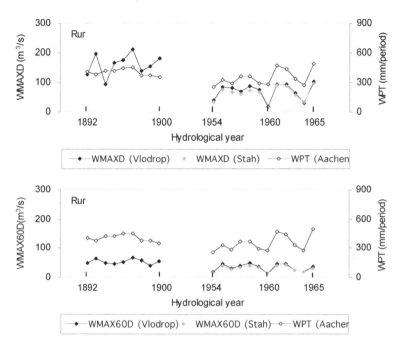

Figure 7.6   Time series of WMAXD and WMAX60D of the Rur (at Vlodrop and Stah) and winter (November–April) precipitation total (WPT, mm/period) at Aachen over two different periods: 1892–1900 and 1954–1965.

(e.g. 40% for WMAXD) in the Geul after 1971 may be explained by the combined effect of the flood mitigating measures (e.g. restoring meanders, retention areas) and the measurement system change (see De Laat and Agor, 2003). Due to the above strong impact, effect of the wetter winter climate over the recent two decades is hidden from the (winter) flood observations. Likewise, the effect of agricultural management practices and urbanisation in the area is not easily observable. The flow regime of the Jeker, as mentioned earlier in section 7.2.2, is less sensitive to seasonal precipitation change and land use change. This feature gives a likely explanation of no evident change in WMAXD of the river after 1981 (in response to the wetter winter climate). The change detected in the day-for-WMAXD series for the Jeker around 1990 is very likely affected by the correcting exercise in the reconstruction of the Nekum record. It should be mentioned that floods in the small rivers Geul and Jeker are more sensitive to summer intense storm events on a single day or over a shorter period. Therefore, attention should be given to their instantaneous peak discharges in summer time. In this respect, one has to recognise the shortcomings of the flood analyses based on the daily discharge data.

7.3.3 Meuse river

It has been shown that the increase of (winter) flood peak discharge in the Meuse river over the recent two decades, along with the increased frequency of (winter) flood events, is mainly a reflection of climate variability. The overall effect of land use changes in the upstream Meuse basin on the (winter) flood trend appears to be marginal or indiscernible. The weak effect is likely due to a combination of the following important facts or reasons:

- Despite a rapid expansion in recent decades, the coverage of the urban areas upstream accounts for only a small fraction (currently about 9%) of the total basin area.

- Moreover, the percentage of the basin covered by forest has not significantly changed in the 20th century (its current coverage being about 35% of the basin area).

- The difference in the antecedent catchment wetness caused by different land use patterns is often small or negligible, compared to the natural variation in the catchment wetness.

- The effects of management practices associated with modern forestry and agriculture in the upstream basin might be observable locally, but would be least significant or indiscernible at the large basin level because of offset effects.

- A large part of the Meuse basin is situated in the hilly Ardennes region with relatively low water storage capacity of soils, accordingly favourable to the formation of floods in response to storm events.

- The pronounced effect of climatic variation is also likely to overshadow the land use effect, making the latter effect less apparent or indiscernible.

The magnitude of winter floods in the Meuse river is more likely to be affected by hydraulic activities in the river network. The possible impacts on the (winter) flood trend mainly come from canal water extraction from the Meuse, retention of the French Meuse valley, regulation of the reservoirs and modification of the river channel. The overall canal extraction rate has been obviously increasing since the end of 1930s (see Figure 2.4). However, in the case of a flood, the total extraction rate is still relatively low, not more than 5% of the natural (annual mean) daily discharge in the Meuse. Neither is it possible to make those canals contribute significantly (some hundreds m^3/s) to the daily discharge of the river, since apart from several constructive problems it is impossible physically, too (Berger, 1992). The influence of the canal extractions on the measurements of flood flow is relatively small.

In the Lorraine Meuse, large volumes of floodwater can be retained in the inundated valley (see Figure 2.19) and, as a result, the flood waves are weakened. In the central Meuse stretch, the flood waves are hardly weakened due to little floodplains available in the narrow and steep Ardennes valley. Most of the reservoirs in the tributaries (upstream of Borgharen) to be used for the flood control purpose and other multiple functions were built after the 1950s. The total storage capacity of the reservoirs and the controlling upstream area (see Figure 2.17) is too small to impact the occurrence of (winter) large floods in the Meuse. Although the effects of the channel modifications on the shape of flood wave and the flood peak in the downstream Meuse river are difficult to assess, they are unlikely to give rise to a considerable increase in the size of winter floods (e.g. more than 400 m^3/s in WMAXD, equivalent to about 30% or more) at Borgharen since 1984.

The occurrence of winter maximum floods (i.e. day-for-WMAXD) in the Meuse was found to have been significantly postponed (on average more than 20 days) after 1942 (see Figure 3.9). One could attempt to make a connection between the postponement and the changes in the forest area and type in the upstream Meuse basin taken place since the 1930s, because higher losses of interception and ET are expected and subsequently tend to increase the soil moisture deficit of the basin after summer dry periods. However, the HBV-simulations indicate that the impact of forest type change is too small to cause a delay in the occurrence of winter maximum floods in the Meuse. The hydraulic activities such as canal extractions, regulation of the reservoirs and utilization of the retention areas are also unlikely to be the main causes. Recall that the winter precipitation extremes of short duration (e.g. WMAX10P) in the Meuse basin have significantly increased after the mid-1930s (see Figure 4.9). This hints at a possible climate-linked cause for the postponement of winter maximum floods in the river. Whether the time of occurrence of WMAX10P in the Meuse basin (based on the MeuseLP record) has significantly changed was investigated. The trend result indeed shows that WMAX10P in the basin occurred with a delay of more than 20 days since 1945 (with a probability of 0.81 for the change point year in the Pettitt test result), which clearly corresponds to the delay observed for the winter maximum floods.

7.4 Effect on the low flows

7.4.1 Introduction

A few modelling studies have shown human impacts on groundwater regimes of some subcatchments in the Meuse basin. In the forementioned case study by Bultot *et al.* (1990), the simulated low flows in the Houille were almost unaffected by land cover change. The modelling results given by Querner (2000) for three lowland catchments in The Netherlands indicate that the natural recharge of groundwater has increased over the last 50 years and the effect of climatic variation was greater than the change caused by human influences. Among the various human intervention scenarios, lowering the drainage base was found to bring about the greatest change in groundwater tables. Van Lanen and Dijksma (1999) also mentioned that relatively small but permanent changes in groundwater recharge perhaps have stronger impact on streamflows of the Geul than the current groundwater abstraction in subcatchment. Based on the simulated results, De Wit *et al.* (2001) reported that the removal of drains and ditches in the lowland Beerze subcatchment might increase the groundwater flow and a complete afforestation of two Belgian subcatchments in the Ardennes might increase the risk of low flows.

   In the following sub-sections, the potential impacts of human activities on summer low flows of the selected tributaries and the Meuse are assessed.

7.4.2 Selected tributaries

Semois river

The striking shift of SMIN10D in the Semois since 1971 can not be explained by climate variation (see Figure 6.13). The Vierre dam was constructed in 1967 (Mergen, 2002; 1970 given by Berger, 1992), close to the occurrence of the shift. However, it is too small to cause such a distinct drop in SMIN10D of the Semois. Neither does the land use change in the subcatchment. The remarkable shift turns out to be mainly caused by the measurement system change at Membre since 1968.

Rur river

The major human activities affecting SMIN10D of the Rur over the second half of the 20th century are large-scale lignite mining and operation of the reservoirs in the upstream part. Figure 7.7 clearly displays an obvious change in SMIN10D of the river after construction of the reservoirs. Relative to the historical period (1892–1900), SMIN10D of the Rur in the latter period (1954–1965) appears to have doubled (5.6 m³/s up to 13 m³/s). The climate data of the two periods for Aachen indicate no significant difference in their winter or summer precipitation amounts (see Figures 7.6 and 7.7). Therefore, the construction of the reservoirs has ultimately led to an overall increase of SMIN10D in the Rur throughout the whole period of record from 1954 to 2001, which is supported by the HBV-simulations in Yusuf (2005). The upward shift in SMIN10D of the Rur around 1965 is very likely to be the consequence of an increasing reservoir capacity after 1960.

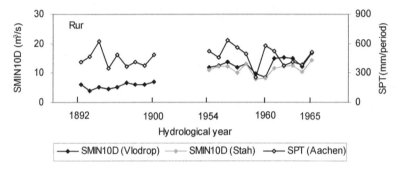

Figure 7.7 Time series of SMIN10D of the Rur (at Vlodrop and Stah) and summer (May–October) precipitation total (SPT, mm/period) at Aachen over two different periods: 1892–1900 and 1954–1965.

The groundwater tables in the open mining pit areas and the adjacent groundwater tables have been lowered due to extraction of groundwater for the mining activities since 1950. One would expect that the lowering of groundwater tables has a significant effect on SMIN10D of the Rur. However, from the analysis results, it seems that any (negative) effect on SMIN10D in the river due to the lowering of groundwater tables from the 1950s, if exists, has been masked by the pronounced (positive) effect of the reservoirs throughout the whole period of record from 1954 to 2001. The decrease of SMIN10D in the Rur after 1988 (see Figure 3.23) is probably climate-linked, similar to the interpretation for the neighbouring rivers Geul and Jeker.

Geul and Jeker rivers

Intensive groundwater abstractions in the Geul and Jeker subcatchments seem to have little effect on variations of SMIN10D in the rivers during their study periods (1954–2001 and 1966–2001, respectively). It is very likely that the hydrological systems of these two subcatchments have been subject to new equilibriums over their whole (or most parts of) record periods. Nevertheless, it is worth mentioning that the magnitudes of SMIN10D in the two rivers might have reduced to some degree as the consequence of long-term groundwater abstractions in the areas.

7.4.3 Meuse river

During low flow periods, the hydraulic works such as canals and weirs in the Meuse (upstream of Borgharen) have substantially affected the discharge downstream. Canal extractions from the Meuse between Liège and Borgharen have been approximately taken into account in the reconstructed Monsin record; hence variations of SMIN10D in the Meuse based on the Monsin record are presumably affected by the canals to a limited degree. The influence of the weirs on the low flow trend of the Meuse (near Monsin) is very small. Figures 3.19 and 3.22 provide some evidences to justify the above assessments. Except for the Eau d'Heure reservoirs on the Sambre, the reservoirs in the upstream Meuse basin are too small to have a significant effect on SMIN10D of the Meuse. Moreover, most of the reservoirs were built in the second half of the 20th century. Therefore, the marked decrease of SMIN10D in the Meuse since 1933 is unlikely due to the influence of the reservoirs.

The drop of SMIN10D in the Meuse around 1933 can not be explained by natural fluctuation of precipitation or PET (see section 6.4). The increase of actual ET loss due to the forest type change in the Meuse basin is very small, as demonstrated by the HBV-simulations (see section 6.5), and hence its impact on groundwater recharge in the forested areas and ultimately SMIN10D of the Meuse is likely to be negligible. No other recognised land use changes in the basin can be connected to the drop.

From the above discussion, interpretation of the decrease in SMIN10D of the Meuse (near Monsin) around 1933 seems to encounter a similar difficulty as previously noted for the decrease of the annual runoff/rainfall ratio around 1932 (section 7.2.3). There may be some influential factor(s) (e.g. quality of the Borgharen discharge record) giving rise to the concurrent changes in both hydrological variables.

7.5 Synthesis

During the 20th century, the most obvious land use changes in the Meuse basin (upstream of Borgharen) are forest type change (from deciduous forest to coniferous forest), modernisation of forestry and agricultural management practices and rapid urbanisation. These types of land use changes have been generally recognised to potentially have the most significant effects on the hydrology regime of a catchment. Recent floods in the Meuse river led to a public perception that the rapid land use changes of recent decades in the upstream part have increased the frequency and magnitude of floods in the downstream Meuse river. However, the in-depth investigation based on both the statistical trend results and the HBV-simulated results clearly demonstrates that variations in the observed discharge record of the Meuse (at Borgharen/Monsin) from 1911 can be largely explained by variations in the meteorological variables (mainly precipitation and PET). The overall impacts of land use changes on the discharge regime of the Meuse appear to be marginal or statistically undetectable. More pronounced are the impacts of hydraulic works and management measures along the river system. The decrease shown in the annual runoff/rainfall ratio of the Meuse basin since the early 1930s can not be explained by the meteorological data (mainly precipitation and PET) and the forest type change. The real cause requires further investigation. The slight increase of the annual runoff/rainfall ratio observed in the last two decades of the 20th century can be reasonably explained by non-linearity in the hydrological response to enhanced precipitation amount. The increased frequency and magnitude of (winter) floods in the Meuse over the last two decades is basically a reflection of climate variability. The results presented in this study do not justify the assumption that land use changes (e.g. deforestation, drainage, road construction, urbanisation) have substantially influenced the occurrence of recent floods in the downstream Meuse river in the last decades. The attenuating effects of hydraulic works and management measures (e.g. reservoirs, retention areas) along the Meuse river system can not impact the occurrence of large floods downstream. The remarkable decrease in the summer low flow of the Meuse (near Monsin) since the early 1930s is most likely related to the quality of the Monsin discharge record analysed. The investigations on the selected subcatchments of varying sizes and different catchment characteristics in the Meuse basin show supportive evidences that the hydrological effects of land use changes in these areas are undetectable within their record periods. The analysis results obtained in this study do support the argument/suspect that the hydrological effects of "non-drastic" land use changes are often insignificant in catchments larger than several hundred km^2 (e.g. Kiersch, 2000; FAO and CIFOR, 2005), and also generally agree with the conclusions of recent investigations on the European large river basins (e.g. Tollan, 2002; Mudelsee *et al.*, 2003; Middelkoop *et al.*, 2004). Several possible reasons may explain why the hydrological effects of land use changes at the catchment level are often hardly detectable: *i*) The extent and intensity of land use changes are still very limited relative to the whole catchment and hence the resulting changes in the hydrological regime are too small to be detected (at an adopted significance level). *ii*) The impacts of multiple land use changes in the whole catchment tend to be balanced out and, as a result, the net effect becomes less obvious. *iii*) The stronger effects of climate variability may overshadow the effects of land use changes, making the latter effects invisible. *iv*) The limited data used do not enable a detailed analysis.

Anthropogenic climate change will persist for many centuries (Houghton et al., 2001). Given the present climate variability and projected climate change, the risk of (winter) flooding in the Meuse river is expected to increase in future (Pfister *et al.*, 2004). New flood management strategies and measures are being developed and formulated. The insights gained by this study allow concluding that at the basin level, the anticipated effects of the "non-drastic" SPONGE-measures and the RETENTION-measures proposed to be implemented in the upstream Meuse basin (see Geilen *et al.*, 2001) may not be noticeable on the downstream part of the river (especially for extreme events) and therefore should not be overestimated, in particular for large floods. To reduce the flood risks, as pointed out by Hooijer *et al.* (2004) in a summative paper, the most effective and sustainable way is to reduce the potential damage (vulnerability) in flood-prone areas through adapted land use and spatial planning.

7.6 Concluding remarks

- The effects of historical land use changes on the discharge regimes of the Meuse river and the selected tributaries are marginal or statistically undetectable.
- More pronounced are the effects of hydraulic works and management measures along the river systems. For example, the construction of the reservoirs in the tributary Rur starting from the early 20th century has led to a significant increase in the summer low flow (SMIN10D) and a significant decrease in the winter flood peak discharge (WMAXD) of the river; the flood mitigating measures along the tributary Geul over recent three decades have substantially reduced WMAXD in the river. However, at the Meuse basin level, the attenuating effects of hydraulic works and management measures along the river network can not impact the occurrence of large floods in the downstream river.

- The decrease of the annual runoff/rainfall ratio for the Meuse since 1932 can not be explained by historical land use changes in the basin. The downward shifts shown in SMIN10D of the Meuse (after 1933) and the Semois (after 1971) are very likely related to the quality of the discharge data analysed.

This chapter aims at getting a general picture of human impacts on the rainfall-runoff relation in the large Meuse basin during the 20th century. Fully quantifying these impacts is not attempted and also not possible. Further researches using a more physically-based modelling approach are suggested, e.g. determining the impacts of hydraulic changes in the Meuse river network, identifying storm runoff contributing areas in the upstream Meuse basin which are most sensitive to land use practices and changes.

8   Summary and conclusions

Potential impacts of climate change/variability on regional (or local) precipitation patterns and, subsequently, the hydrology of individual river basins have received a growing attention. This research concentrates on the Meuse river, one of the largest river basins in northwestern Europe. The Meuse river is an almost entirely rain-fed river and has a discharge regime characterised by floods occurring during winter and low flows occurring during summer and autumn. Recent floods in the Meuse river have increased the public's concern about the risk of flooding and have also stimulated discussions about the possible effects of climate variability/change and land use change in the river basin. This research contributes to a better understanding of the hydrological response of the Meuse river basin to climate variability (preferably used in the present thesis) and land use change.

With available long and reliable daily observation records (starting from 1911), the temporal changes in the discharge and precipitation regimes in the Meuse basin upstream of Borgharen (about 21,260 km^2, see Figure 2.2) have been investigated using statistical trend detection methods. To obtain supportive evidences and also to demonstrate effect of spatial scale, the investigation was extended to a few selected tributaries (i.e. Rur, Semois, Jeker and Geul) with varying catchment sizes (338–2,245 km^2) and different catchment characteristics, for which the relatively long daily discharge records (starting from 1929 and 1953) were available or had been reconstructed. Furthermore, a synoptic-climatological analysis has been carried out to understand the observed precipitation changes in the Meuse basin. The rainfall-runoff relations in the study areas (emphasizing the entire Meuse basin) have been statistically analysed with reference to different flow states (i.e. mean flows, high flows and low flows), complemented by a hydrological modelling study (using the semi-distributed conceptual HBV model) for the entire Meuse basin to gain further insights into the effects of climate variation and the historical change in the forest type. In addition, potential evapotranspiration in the Meuse basin has also been statistically analysed and included in the assessment of the relations. Historical land use changes and hydraulic activities in the Meuse basin occurring in the 20th century have been identified through a variety of data sources and their potential effects on the discharge regime of the downstream Meuse river have been qualitatively assessed. In the statistical trend analyses, the identification of abrupt changes or shifts (regarding the mean values) in the observation series (using both the non-parametric Pettitt test and the parametric SNHT test) was emphasized, which is very useful when searching for possible reasons behind the changing hydrological behaviour of the river system. The trend detection exercise demonstrated that the Pettitt test is powerful to detect subtle changes in the observation series. Comparatively, the detection of linear trends (using the non-parametric Spearman's rank test) receives less attention. Due to relatively high variability of the observation series, the results of the Spearman test may easily suggest absence of significant linear trends. In the interpretations, the shortcomings of the observation records used and the limitations of the methodologies applied have been taken into account. In the following parts, the main results obtained from this research (focusing on the entire Meuse basin) are summarized and then the conclusions and recommendations for further research are given.

Changing precipitation regime and its link to synoptic climatology

Trend analysis of the basin-averaged precipitation record (estimated from seven Belgian gauging stations, 1911–2002) for the Meuse basin reveals that there exist (apparent) changes in the precipitation amounts, precipitation events and precipitation extremes during the 20th century and most of the observed changes occurred in the last two decades. The annual precipitation amount (November–October) in the basin appears to have just significantly increased since 1980, accompanied by some varying changes in the seasons, e.g. a wetter (extended) winter (December–March) since 1979 and a slightly drier summer (June–August, mainly August) since 1967. There is evidence for more intense precipitation events (e.g. very wet days with daily areal precipitation exceeding 10 mm/d) in the Meuse basin occurring in the winter half-year (November–April) since 1980. The change in precipitation pattern is likely (but not necessary) to have had effect on the precipitation extremes of certain durations (e.g. ranging from one day to 30 days) in the Meuse basin. The tendency to increased precipitation amounts and increased precipitation intensities in winter in the Meuse basin is in line with the regional precipitation trend observed in northern Europe (e.g. Folland *et al.*, 2001).

The precipitation pattern change in the Meuse basin since 1980 has been found to reflect the fluctuation of large-scale atmospheric circulation, as characterised by the subjective classification system European atmospheric circulation patterns according to Hess and Brezowsky (*Grosswetterlagen*; see Gerstengarbe and Werner, 1999). In the winter half-year, the frequencies of wet days and very wet days (with daily areal precipitation exceeding 1 mm/d and 10 mm/d, respectively) in the Meuse basin contributed by the rain-associated circulation patterns (e.g. West cyclonic Wz, Southwest cyclonic SWz and Northwest cyclonic NWz) as well as their associated precipitation amounts have increased considerably since 1980. The winter (December–February) precipitation in the Meuse basin shows a positive correlation (significant at 1% level) with the winter (December–February averaged) NAO index (the Gibraltar-minus-Iceland version; see Jones *et al.*, 1997) over the study period. This implies that the tendency to more abundant precipitation in winter in the Meuse basin since 1980 (in the last several decades) is likely to be a consequence of the strengthened North Atlantic Oscillation (NAO) that brings stronger westerly surface winds across the North Atlantic to Europe.

Changing discharge regime in the context of climate variability

Based on the reconstructed discharge record (1911–2000) for the "undivided" Meuse near Monsin (see Figure 2.2), the annual average discharge appears to have been relatively stable during the 20th century. However, analysis of the annual runoff/rainfall ratio suggests that the annual rainfall-runoff relation in the Meuse basin has changed around 1932, causing a significant decrease in the runoff proportion. There is a difficulty in explaining such a change on the basis of precipitation variability in the area. Annual potential evapotranspiration in the Meuse basin, estimated at a Dutch meteorological station De Bilt in the same climatic region, shows climate-induced fluctuations with enhanced rates during the 1930s to the 1940s and the 1990s. The HBV-simulations of the daily discharge record for Monsin (1911–2000) demonstrated that the decrease of the runoff proportion around 1932 can not be explained by the meteorological data (precipitation, temperature and potential evapotranspiration) used, the forest type change in the basin and the uncertainty of the discharge record for Borgharen/Monsin.

The flood trend of the Meuse river has been investigated based on both annual/winter maximum (AM/WM) series and peaks-over-threshold (POT) series (with three different thresholds) derived from the daily discharge record (1911–2002) for Borgharen (see Figure 2.2). The trend results of two types of flood magnitude series are largely consistent, both (AM/WM series vs. POT series with two different thresholds of 800 m^3/s and 1217 m^3/s) demonstrating that the magnitude of winter (November–April) floods in the Meuse has significantly increased since 1984. Simultaneously, the trend results of the POT frequency series suggest an increased frequency of winter floods (e.g. those greater than 800 m^3/s) in the river since 1979, with more small floods (e.g. those smaller than 1200 m^3/s) occurring in spring (March–May). In addition, the timing of occurrence of winter maximum floods in the Meuse has apparently postponed (about 20 days) since 1942. The increased magnitude of winter floods in the Meuse river over the last couple of decades is largely affected by increased antecedent precipitation depth over consecutive days (e.g. ten days), and thus can broadly be ascribed to climate variability. Supportive evidences have been observed in the selected tributaries (excluding the Jeker and the Geul), which show concurrent increases (statistically significant) in the magnitudes of winter floods during the investigated periods. Furthermore, the HBV-simulations demonstrated the dominating role of meteorological conditions (mainly precipitation) in the flood generation in the Meuse basin. The climate-induced, increasing (winter) flood trend in the large Meuse basin over the last two decades agrees with the findings of many European studies in recent literature.

Based on the reconstructed Monsin discharge record (1911–2000), the summer low flow of the Meuse river, defined as the summer (May–October) minimum consecutive 10-day moving average discharge, shows a marked decrease since 1933. Graphical analysis indicates that except for the distinct downward shift, variations in the summer flow of the Meuse river generally correspond to variations in the meteorological conditions (mainly precipitation and potential evapotranspiration). In spite of some seasonal precipitation changes (i.e. wetter winter and drier summer) occurring in the last two to three decades, there is no clear indication that the magnitude of the summer low flow in the Meuse river has been significantly affected. However, on a local scale, the effect of seasonal precipitation changes depends very much on the storage capacity of the catchment. For example, the summer low flow of the Semois (situated in the area with low water storage capacity) has decreased (to some extent) since 1971, whereas the summer low flows of the Geul and Jeker rivers (situated in the areas with large groundwater storage capacity) remain relatively stable. Finally, it must be stressed that the summer low flow of the Meuse river is sensitive to the influences of many factors, e.g. measurement errors, operation of weirs in the river for shipping, the shortcomings of the reconstructed discharge record etc., and thus is associated with considerable uncertainty. The marked decrease in the summer low flow of the Meuse river around 1933 is likely related to the quality of the Borgharen/Monsin discharge record analysed.

Historical land use changes and hydraulic activities, and the potential impacts

During the 20th century, the most obvious land use changes in the Meuse basin (upstream of Borgharen) are forest type change (from deciduous forest to coniferous forest since the 1920s), modernisation of forestry and agricultural management practices as well as rapid urbanisation (since the 1950s). In addition, many hydraulic engineering works and management measures have been taken place in the Meuse river network, e.g. operation of weirs for navigation, canalisation and water abstraction from the Meuse, construction of the reservoirs, modification of the river channel etc.

The in-depth investigation based on both results of the statistical trend analysis and the HBV-simulations clearly demonstrated that variations in the observed discharge record of the Meuse river (at Borgharen/Monsin) from 1911 onwards can be largely explained by variations in the meteorological variables (mainly precipitation and potential evapotranspiration). The overall effects of historical land use changes in the basin on the discharge regime of the Meuse river are marginal or statistically undetectable. The effects of hydraulic works can be observed locally, e.g. in the Rur river where the reservoirs in the upper part have led to a significant increase in the summer low flow and a significant decrease in the magnitude of winter floods in the river. However, at the Meuse basin level, the attenuating effects of hydraulic works and management measures along the river network are too small to significantly impact the long-time trend of flood discharges downstream. The results of the investigation do not justify the assumption that land use changes upstream have substantially influenced the occurrence of recent floods in the downstream Meuse river; rather, do support the argument/suspect that the hydrological effects, in particular to large floods, of land use changes are often insignificant in catchments larger than several hundred km^2 (e.g. Kiersch, 2000), and also generally agree with the findings of recent investigations in the European large river basins.

Main conclusions and recommendations

The precipitation pattern in the Meuse basin has significantly changed since 1980, characterised primarily by a tendency towards more frequent intense precipitation events and increased precipitation amount in the winter time (November–April). The increased magnitude and frequency of winter floods in the Meuse river over the last two decades of the 20$^\text{th}$ century is basically a reflection of climate variability. There is no evidence that changes in the physical properties of the Meuse river basin, which have taken place during the last century, have significantly affected the occurrence of extreme floods in the Meuse. Figure 8.1 summarizes the effects of climate variability and land use change on the hydrology of the Meuse river basin.

To improve our understanding of human impacts at the Meuse basin level, some further researches are recommended, including: *i*) a detailed analysis of the climatic input and the rainfall-runoff relation in the neighbouring Moselle basin during the 20$^\text{th}$ century, which may help to understand the real cause of the decrease in the annual runoff/rainfall ratio of the Meuse basin since 1932; *ii*) a modelling study on how large-scale hydraulic changes in the river network affect the high flows of the Meuse, futher developing the approach taken by Lammersen *et al.* (2002) for the Rhine; *iii*) a detailed process-based hydrological modelling study combined with experimental investigations using state of the art tracer and geophysical methods, on identifying storm runoff contributing areas in the upstream Meuse basin which are most sensitive to land use practices. The last approach would enable investigations of the impact of land use changes on hydrological processes at different scales (headwater catchment vs. meso-scale catchment vs. river basin scale).

Summary and Conclusions

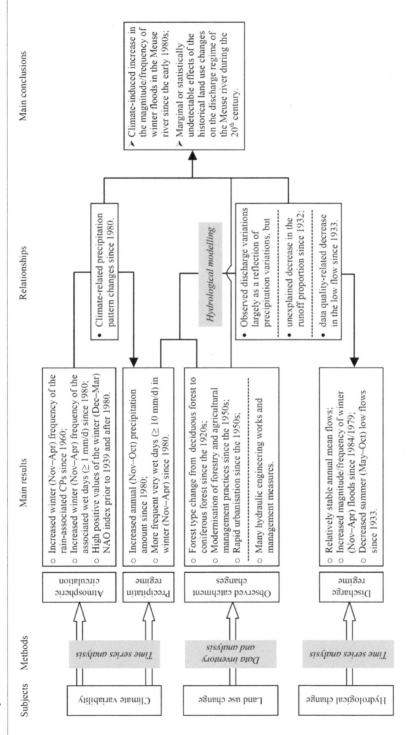

Figure 8.1 Assessment of the effects of climate variability and land use change on the hydrology of the Meuse river basin (upstream of Borgharen) during the 20th century. CP = (atmospheric) circulation pattern.

Samenvatting (Summary in Dutch)

Mogelijke effecten van klimaatsverandering/-variabiliteit op het afvoerregime van rivieren staat in toenemende mate in de belangstelling. Dit onderzoek spits zich toe op het stroomgebied van de Maas, een van de grootste stroomgebieden in noordwest Europa. De Maas is een regenrivier en het afvoerregime wordt gekarakteriseerd door hoogwaters in de winter, en lage afvoeren in de zomer en het najaar. Recente hoogwaters in de Maas hebben tot verhoogde bezorgdheid geleid bij het publiek met betrekking tot het overstromingsrisico en heeft een discussie op gang gebracht over de mogelijke effecten van klimaatsverandering/-variabiliteit en verandering in het landgebruik in het stroomgebied. Dit onderzoek draagt bij tot een beter begrip van de gevolgen van klimaatsvariabiliteit (deze term wordt bij voorkeur in dit proefschrift gebruikt) en verandering van het landgebruik voor de hydrologie van de Maas.

Lange, betrouwbare tijdreeksen van dagwaarnemingen (vanaf 1911) van afvoer en neerslag in het stroomgebied van de Maas bovenstrooms van Borgharen (ongeveer 21260 km^2, zie figuur 2.2) zijn statistisch onderzocht met behulp van trend onderzoekstechnieken op het voorkomen van tijdsafhankelijke veranderingen. Ter ondersteuning van het onderzoek, maar ook om het effect van ruimtelijke schaal te bestuderen, werd het onderzoek uitgebreid met een paar zijrivieren (m.n. de Roer, de Semois, de Jeker en de Geul) met uiteenlopende afmetingen (338–2245 km^2) waarvoor relatief lange dagwaarnemingsreeksen (vanaf 1929 en 1959) beschikbaar waren of gereconstrueerd konden worden. Verder is een synoptisch-klimatologische analyse uitgevoerd voor een beter begrip van de waargenomen verandering in het neerslagregime in het stroomgebied van de Maas. De neerslag-afvoer relaties in de studiegebieden (waarbij de nadruk is gelegd op het gehele stroomgebied van de Maas) werden statistisch geanalyseerd met betrekking tot verschillende afvoerparameters (m.n. hoogwater, en gemiddelde- en lage afvoer). Ter aanvulling werd een hydrologische studie uitgevoerd (met het semi-gedistribueerde conceptuele HBV model) voor het gehele stroomgebied van de Maas om meer inzicht te krijgen in de effecten van klimaatsvariatie en de historische verandering van het bostype. Verder werd de potentiële evapotranspiratie in het stroomgebied van de Maas statistisch geanalyseerd en bij de beoordeling van de effecten betrokken. Er is een overzicht gemaakt van historische veranderingen in het landgebruik en waterbouwkundige activiteiten in het stroomgebied van de Maas gedurende de 20^e eeuw. De mogelijke effecten van deze veranderingen op het afvoerregime van het benedenstroomse deel van de Maas werden kwalitatief beoordeeld. In de statistische trend analyses wordt de identificatie van plotselinge veranderingen of sprongen (met betrekking tot de gemiddelde waarden) in de waarnemingsreeks benadrukt. Hierbij is gebruik gemaakt van de niet-parametrische Pettitt test en de parametrische SNHT test. Deze aanpak is bijzonder waardevol bij het zoeken naar mogelijke oorzaken van het veranderende hydrologische gedrag van het riviersysteem. Gebleken is dat de Pettitt test een krachtig middel is om subtiele veranderingen in de waarnemingsreeks op te sporen. Vergeleken hiermee is er minder aandacht besteed aan de test voor lineaire trends (de niet-parametrische Spearman's test), omdat, vanwege de relatief grote fluctuaties in de waarnemingsreeks, de resultaten hiervan gemakkelijk de suggestie kunnen wekken dat er geen sprake is van significante lineaire trends. De tekortkomingen van de gebruikte waarnemingsreeksen en de beperkingen van de toegepaste methoden zijn bij de interpretatie van de resultaten in beschouwing genomen. De belangrijkste onderzoeksresultaten (met de nadruk op het gehele stroomgebied van de Maas) en de aanbevelingen voor verder onderzoek zijn hieronder samengevat.

Het veranderend neerslagregime en zijn relatie tot de synoptische klimatologie

Trendanalyse van de gebiedsgemiddelde neerslagreeks voor de Maas (gebaseerd op zeven Belgische neerslagstations over de periode 1911–2002) laat zien dat er in de 20^e eeuw (schijnbare) veranderingen zijn optreden in de neerslaghoeveelheden, de neerslagbuien en de neerslagextremen en dat de meeste veranderingen plaats vonden in de laatste twee decennia. De jaarlijkse neerslagsom (november–oktober) in het stroomgebied blijkt sinds 1980 statistisch significant te zijn toegenomen. De seizoensneerslag laat een kenmerkende toename zien in de verlengde winterperiode (december–maart) sinds 1979 en een lichte afname in de zomerperiode (juni–augustus, maar vooral in augustus) sinds 1967. Aangetoond is dat er sinds 1980 in het winterhalfjaar (november–april) meer intensieve neerslagbuien (m.n. erg natte dagen met gebiedsgemiddelde dagneerslag van meer dan 10 mm/d) in het stroomgebied van de Maas voorkomen. Het ligt voor de hand (maar dat is niet zeker) dat de verandering van het neerslagpatroon in het stroomgebied van de Maas effect heeft gehad op neerslagextremen van een bepaalde duur (m.n. die tussen 1 en 30 dagen). De tendens van toegenomen neerslaghoeveelheden en -intensiteiten in het stroomgebied van de Maas komt overeen met de regionale neerslagtrend zoals waargenomen in noord Europa (b.v. Folland *et al.*, 2001).

Geconstateerd is dat de verandering van het neerslagpatroon in het stroomgebied van de Maas sinds 1980 een reflectie is van de fluctuatie van de grootschalige atmosferische circulatie, zoals gekarakteriseerd door het subjectieve classificatiesysteem van Europese atmosferische circulatiepatronen volgens Hess en Brezovsky (*Grosswetterlagen*; zie Gerstengarbe en Werner, 1999). De frequentie van natte dagen en zeer natte dagen in het stroomgebied van de Maas (met een gebiedsgemiddelde dagneerslag van meer dan, respectievelijk 1 en 10 mm/d), gevoed door neerslag-gerelateerde circulatiepatronen (zoals de West cyclonaal Wz, Zuidwest cyclonaal SWz en Noordwest cyclonaal NWz) zijn sinds 1980, evenals de daarmee samenhangende neerslaghoeveelheden, in het winter halfjaar aanzienlijk toegenomen. De neerslag in het stroomgebied van de Maas in het winterseizoen (december–februari) laat, voor het tijdvak van deze studie, een positieve correlatie zien (met een significantie van 1%) met de winter (december–februari gemiddelde) NAO index (de Gibraltar-minus-IJsland versie; zie Jones *et al.*, 1997). Dit impliceert dat de tendens tot meer overvloedige neerslag in de winter in het stroomgebied sinds 1980 waarschijnlijk een gevolg is van de versterkte North Atlantic Oscillation (NAO) die sterkere westenwinden over de Noord Atlantische oceaan naar Europa brengt.

Het veranderende afvoerregime in de context van klimaatsvariabiliteit

Uit de gereconstrueerde afvoerreeks (1911–2000) van de "ongedeelde" Maas nabij Monsin (zie figuur 2.2) blijkt dat gedurende de 20^e eeuw de gemiddelde jaarlijkse afvoer nauwelijks is veranderd. Analyse van het jaarlijkse afvoer/neerslagquotiënt suggereert echter dat de jaarlijkse neerslag-afvoerrelatie in het stroomgebied rond 1932 is veranderd, resulterend in een significante afname van de relatieve afvoer.

Deze verandering kan moeilijk aan de neerslagvariabiliteit worden toegeschreven. De jaarlijkse potentiële evapotranspiratie in het stroomgebied van de Maas berekend met meteorologische gegevens van het Nederlandse meteorologische station in De Bilt dat in hetzelfde klimatologische gebied is gelegen, laat een klimaat-geïnduceerde fluctuatie zien met verhoogde waarden in de dertiger en veertiger jaren, en in de negentiger jaren. Simulaties van de afvoerreeks voor Monsin op dagbasis met het HBV model tonen aan dat de afname van de relatieve afvoer rond 1932 niet verklaard kan worden door de meteorologische gegevens (neerslag, temperatuur en potentiële evapotranspiratie), de verandering van het bostype in het stroomgebied en de onnauwkeurigheid van de afvoerreeks voor Borgharen/Monsin.

De hoogwatertrend voor de Maas werd onderzocht voor zowel de reeks jaarlijkse- als winter maxima (AM/WM) en voor de van de afvoerreeks van Borgharen (1911–2002) afgeleide partiële series (POT, Peaks-Over-Threshold) voor drie verschillende drempelwaarden (zie figuur 2.2). De resultaten van de analyses van de twee typen hoogwaterreeksen komen grotendeels overeen. Met beide reeksen (de AM/WM series en de partiële series met twee verschillende drempelwaarden van 800 en 1217 m^3/s) wordt aangetoond dat het winterhoogwater (november–april) in de Maas significant is toegenomen sinds 1984. Tegelijkertijd suggereren de trendresultaten van de partiële series een toegenomen frequentie sinds 1979 van extreme afvoeren in de winter (m.n. die groter dan 800 m^3/s) en het meer frequent voorkomen van geringe hoogwaters in de lente (maart–mei). Verder blijkt dat de wintermaxima in de Maas zich sinds 1942 later voordoen (ongeveer 20 dagen). De toegenomen grootte van de piekafvoer van de Maas in de winter gedurende de laatste paar decennia is voornamelijk het gevolg van de toegenomen neerslag over de voorafgaande (tien) dagen en kan dus goeddeels worden toegeschreven aan klimaatsvariabiliteit. Deze resultaten werden bevestigd voor waarnemingen in de geselecteerde zijrivieren (behalve voor de Geul), die een gelijktijdige toename laten zien (zowel significant als niet-significant) in de grootte van de piekafvoer in de winter voor de onderzochte perioden. Verder lieten de HBV simulaties de dominante rol zien van de meteorologische omstandigheden (voornamelijk neerslag) in het genereren van hoogwater in de Maas. De klimaat-geïnduceerde trend die een toename laat zien van hoogwater in de winter in het grote stroomgebied van de Maas over de laatste twee decennia komt goed overeen met de uitkomsten van menig Europese studie in de recente literatuur.

De gereconstrueerde afvoerreeks van Monsin (1911–2000) laat zien dat de lage afvoer van de Maas in de zomer merkbaar is afgenomen sinds 1933, waarbij de lage afvoer is gedefinieerd als het minimum in de zomer (mei–oktober) van het voortgaande gemiddelde van 10 opeenvolgende dagen. Een grafische analyse geeft de indruk dat, behalve voor de duidelijk neerwaartse beweging, variaties in de lage afvoer van de Maas in de zomer over het algemeen corresponderen met variaties in de meteorologische omstandigheden (voornamelijk neerslag en potentiële evapotranspiratie). Ondanks enige veranderingen in de seizoensneerslag (t.w. een nattere winter en een enigszins drogere zomer) zoals waargenomen in de laatste twee tot drie decennia, is er geen duidelijke aanwijzing dat de grootte van de lage afvoeren van de Maas in de zomer significant zijn beïnvloed. Op lokale schaal hangt het effect van veranderingen in de seizoensneerslag echter zeer sterk af van de bergingcapaciteit van het stroomgebied. Bijvoorbeeld, de lage afvoer in de zomer van de Semois (gesitueerd in een gebied met geringe bergingscapaciteit) is (tot op zekere hoogte) afgenomen sinds 1971, terwijl de lage afvoer in de zomer van de Geul en de Jeker (gesitueerd in gebieden met grote grondwaterbergingscapaciteit) relatief stabiel zijn gebleven. Tenslotte moet worden benadrukt dat de lage afvoer van de Maas in de zomer door verschillende factoren wordt beïnvloed, zoals meetfouten, beheer van stuwen in de rivier voor navigatie, tekortkomingen bij de reconstructie van de afvoerreeks, enz., en dus behept met een aanzienlijke onzekerheid. De merkbare afname van de lage afvoer van de Maas sinds begin jaren dertig is waarschijnlijk te wijten aan de kwaliteit van de geanalyseerde afvoerreeks van Borgharen/Monsin.

Historische veranderingen in het landgebruik en waterbouwkundige activiteiten, en de potentiële gevolgen

De meest opvallende veranderingen in het landgebruik in het stroomgebied van de Maas (bovenstrooms van Borgharen) zijn de verandering van bostype (van loofbos naar naaldbos sinds de twintiger jaren), de modernisering van de bos- en landbouw alsmede de snelle urbanisatie (sinds de vijftiger jaren). Bovendien hebben vele waterbouwkundige werken en beheersmaatregelen in de riviertakken van de Maas plaats gevonden, zoals het beheer van stuwen ten behoeve van de scheepvaart, kanalisatie en wateronttrekking aan de Maas, de bouw van reservoirs, verandering van de stroomgeul, enz.

Het grondige onderzoek, gebaseerd op zowel de resultaten van statistische trend analyses en de HBV modelsimulaties heeft duidelijk aangetoond dat variaties in de waargenomen afvoerreeks van de Maas bij Borgharen/Monsin vanaf 1911 grotendeels verklaard kunnen worden uit variaties in de meteorologische variabelen (voornamelijk neerslag en evapotranspiratie). Globale effecten van het historisch veranderd landgebruik op het afvoerregime van de Maas zijn marginaal of statistisch niet waarneembaar. De effecten van waterbouwkundige werken kunnen alleen lokaal worden waargenomen, zoals in de Roer waar reservoirs in het bovenste deel van het stroomgebied hebben geleid tot een merkbare toename van de lage afvoer in de zomer en een significante afname van de piekafvoer in de winter. Voor het gehele stroomgebied van de Maas zijn de nivellerende effecten van de waterbouwkundige werken en beheersmaatregelen in de riviertakken echter te gering om een significante uitwerking te hebben op de lange-termijn trend van de benedenstroomse piekafvoer. De resultaten van het onderzoek rechtvaardigen niet de aanname dat bovenstroomse veranderingen in het landgebruik de benedenstroomse hoogwaters van de Maas wezenlijk hebben beïnvloed, maar ondersteunen eerder het argument/vermoeden dat de hydrologische effecten van veranderingen in het landgebruik, met betrekking tot m.n. hoge afvoerpieken dikwijls niet significant zijn in stroomgebieden groter dan enige honderden km2 (b.v. Kiersch, 2000), hetgeen over het algemeen ook overeenstemt met de uitkomsten van recente onderzoeken in de grote Europese stroomgebieden.

Voornaamste conclusies en aanbevelingen

Het neerslagpatroon in het stroomgebied van de Maas is sinds 1980 significant veranderd. Deze verandering wordt voornamelijk gekarakteriseerd door een tendens naar meer frequente intensieve buien en toegenomen neerslaghoeveelheden in de winter periode (november–april). De toegenomen grootte en frequentie van de piekafvoer van de Maas in de winter gedurende de laatste twee decennia van de 20^e eeuw is op zich een reflectie van de klimaatsvariabiliteit. Het bleek niet mogelijk om aan te tonen dat de fysieke veranderingen in het stroomgebied van de Maas gedurende de vorige eeuw van invloed zijn geweest op de toename van de hoogwaterfrequentie.

Om ons begrip van de gevolgen van menselijk ingrijpen in het gehele stroomgebied van de Maas te verbeteren, worden de volgende nadere onderzoeken aanbevolen: *i*) een gedetailleerde analyse van de klimatologische gegevens en de neerslag-afvoerrelatie van het naburige stroomgebied van de Moesel gedurende de 20^e eeuw, omdat dit een bijdrage kan leveren aan een beter begrip van de werkelijke oorzaak van de afname van het jaarlijkse afvoer/neerslagquotiënt van het stroomgebied van de Maas sinds 1932; *ii*) een modelstudie over hoe grootschalige waterbouwkundige veranderingen in de riviertakken de hoge afvoeren van de Maas beïnvloeden, voortbouwend op de aanpak van Lammersen *et al.*, (2002) voor de Rijn; *iii*) een gedetailleerde, fysisch gebaseerde hydrologische modelstudie gecombineerd met experimenteel onderzoek met gebruikmaking van de laatste tracer- en geofysische methoden om afvoer genererende gebieden in het bovenstroomse deel van de Maas te identificeren die het meest gevoelig zijn voor grond- en landgebruik. Deze laatste benadering zou het mogelijk maken de gevolgen van verandering in het landgebruik op de hydrologische processen op verschillende schaalniveaus (bovenloop versus middenloop versus gehele stroomgebied) te onderzoeken.

References

Agor, M.L.C., 2003. Assessment of the long-term rainfall-runoff relation of the Geul catchment. MSc Thesis, UNESCO-IHE, Delft, The Netherlands.

Alexander, L.V., Zhang, X., Peterson, T.C. *et al.*, 2006. Global observed changes in daily climate extremes of temperature and precipitation. J. Geophys. Res., 111, D05109, doi:10.1029/2005JD006290.

Alexandersson, H., 1986. A homogeneity test applied to precipitation data. J. Climatol., 6(6), 661–675.

Alexandersson, H. and Moberg, A., 1997. Homogenization of Swedish temperature data, Part I: Homogeneity test for linear trends. Int. J. Climatol., 17(1), 25–34.

Allen, R.G., Pereira, L.S., Raes, D. and Smith, M., 1998. Crop evapotranspiration – guidelines for computing crop water requirements, Annex 4 Statistical analysis of weather data sets, FAO Irrigation and Drainage Paper no. 56, FAO, Rome, 229–243.

Arends, M., 2005. Low flow modelling of the Meuse: recalibration of the existing HBV Meuse model. MSc Thesis, University Twente, Enschede, The Netherlands. http://www.wem.ctw.utwente.nl/onderwijs/afstuderen/afstudeerverslagen/2005

Arknature, 2000. Flowing storage: a nature solution for high-water problems along brooks and rivers. http://www.arknature.org/[2002. August 8]

Ashagrie, A.G., 2005. Assessment of the impact of land use change on the rainfall-runoff relation of the Meuse river. MSc Thesis, UNESCO-IHE, Delft, The Netherlands.

Ashagrie, A.G., De Laat, P.J.M., De Wit, M.J.M., Tu, M. and Uhlenbrook, S., 2006. Detecting the influence of land use changes on Floods in the Meuse River Basin: the predictive power of a ninety-year rainfall-runoff relation. Hydrol. Earth Syst. Sci. Discuss., 3, 529–559.

Bárdossy A. and Caspary H.J., 1990. Detection of climate change in Europe by analysing European atmospheric circulation patterns from 1881 to 1989. Theor. Appl. Climatol., 42(3–4), 155–167.

Baulies, X. and Szejwach, G. (Eds.), 1997. LUCC data requirements workshop: survey of needs, gaps and priorities on data for land-use/land-cover change research. LUCC Report Series no.3.

Beighley, R.E. and Moglen, G.E., 2002. Trend assessment in rainfall-runoff behavior in urbanizing watersheds. J. Hydrol. Eng., 7, 27–34.

Berger, H.E.J., 1992. Flow forecasting for the river Meuse. PhD Thesis, Delft University of Technology, Delft, The Netherlands.

Bergström, S. and Forsman, A., 1973. Development of a conceptual deterministic rainfall-runoff model. Nord. Hydrol., 4, 147–170.

Beven, K.J., 1996a. A discussion of distributed hydrological modelling. In: M.B. Abbott and J.C. Refsgaard (Eds.), Distributed hydrological modelling, Kluwer Academic, Dordrecht, The Netherlands, 255–278.

Beven, K.J., 1996b. Response to comments on "A discussion of distributed hydrological modelling" by J.C. Refsgaard *et al.* In: M.B. Abbott and J.C. Refsgaard (Eds.), Distributed hydrological modelling, Kluwer Academic, Dordrecht, The Netherlands, 289–295.

Beven, K.J., 2001a. How far can we go in distributed hydrological modelling? Hydrol. Earth Sys. Sci., 5, 1–12.

Beven, K.J, 2001b. Rainfall-runoff modelling: the primer. John Wiley & Sons Ltd., England, UK.

Black, P.E., 1991. Watershed hydrology. Prentice Hall Inc., New Jersey, USA.

Blöschl, G. and Sivapalan, M., 1995. Scale issues in hydrological modelling: a review. Hydrol. Process., 9(3–4), 251–290.

Bonell, M., 1998. Selected challenges in runoff generation research in forests from the hillslope to headwater drainage basin scale. J. Am. Water Res. Assoc., 34, 765–786.

Booij, M.J., 2002. Appropriate modelling of climate change impacts on river flooding. PhD Thesis, University Twente, Enschede, The Netherlands.

Booij, M.J., 2005. Impact of climate change on river flooding assessed with different spatial model resolutions. J. Hydrol., 303, 176–198.

Bos, H., 1993. Verloop daggemiddelde afvoer Borgharen, periode 1911–1991. RIZA werkdocument 92.112X. RIZA, Lelystad, The Netherlands (in Dutch).

Bosch, J.M. and Hewlett, J.D., 1982. A review of catchment experiments to determine the effect of vegetation changes on water yield evapotranspiration. J. Hydrol., 55, 3–23.

Bouwer L.M., 2001. Climate change and the rising cost of river flooding. MSc Thesis, Faculty of Earth Sciences, Vrije Universiteit Amsterdam, The Netherlands.

BráZdil, R., Pfister, C., Wanner, H., Von Storch, H. and Luterbacher, J., 2005. Historical climatology in Europe – the state of the art. Clim. Chan., 70(3), 363–430.

Bronstert, A., 1995. River flooding in Germany: influenced by climate change? Phys. Chem. Earth, 20(5–6), 445–450.

Bronstert, A., Vollmer, S. and Ihringer, J., 1995. A review of the impact of land consolidation on runoff production and flooding in Germany, Phys. Chem. Earth, 20(3–4), 321–329.

Bronstert, A., Niehoff, D. and Burger, G., 2002. Effects of climate and landuse change on storm runoff generation: present knowledge and modelling capabilities. Hydrol. Process., 16, 509–529.

Brooks, K.N., Folliott, P.F., Gregersen, H.M. and De Bano, L.F., 1997. Hydrology and the management of watersheds (2nd ed). Iowa State University Press.

Brouyère, S., Carabin, G., and Dassargues, A., 2004. Climate change impacts on groundwater resources: modelled deficits in a chalky aquifer, Geer basin, Belgium. Hydrogeol. J., 12(2), 123–134.

Buchtele, J., 1993. Runoff changes simulated using a rainfall-runoff model. Water Resour. Manage., 7(4), 273–287.

Bultot, F., Dupriez, G.L. and Gellens, D., 1990. Simulation of land use changes and impacts on the water balance: a case study for Belgium. J. Hydrol., 114(3–4), 327–348.

BUND, 2002. Lignite mining in the Rhineland. Bund für Umwelt und Naturschutz Deutschland, Landesverbände Nordrhein-Westfallen, Düsseldorf, Germany. http://www.bund-nrw.org/bund-braunkohlen-ausstiegs-szenario.htm 12/20/2004.

Burn, D.H. and Hag Elnur, M.A., 2002. Detection of hydrological trends and variability. J. Hydrol. 255, 107–122.

Calder, I.R., 1993. Chapter 13 Hydrologic effects of land-use change. In: D.R. Maidment, (Ed.), Handbook of hydrology, McGraw-Hill Inc., 13.1–13.50, New York.

Calder, I.R., 1998. Water-resource and land-use issues. SWIM Paper 3, International Water Management Institute, Colombo, Sri Lanka.
http://www.cgiar.org/iwmi/pubs/swimpubs/Swim03.pdf

Calderón Hijuma, M.P., 2003. Assessment of the long-term rainfall runoff relation of the Rur catchment. Internal report, UNESCO-IHE, Delft, The Netherlands.

Caspary, H.J., 1995. Recent winter floods in Germany caused by changes in the atmospheric circulation across Europe, Phys. Chem. Earth, 20 (5–6), 459–462.

Cavadias, G.S., 1995. Climatic change, river flows and water resources development. In: G.W. Kite (Ed.), Time and the river, Essays by Eminent Hydrologists, Water Resources Publications, Colorado, USA, 259–272.

Chbab, E.H., 1995. How extreme were the 1995 flood waves on the rivers Rhine and Meuse? Phys. Chem. Earth, 20(5–6), 455–458.

Conley, L.C. and McCuen, R.H., 1997. Modified critical values for Spearman's test of serial correlation. J. Hydrol. Eng., 2, 133–135.

Corti, S., Molteni, F. and Palmer, T.N., 1999. Signature of recent climate change in frequencies of natural atmospheric circulation regimes. Nature, 398(6730), 799–802.

Dahmen, E.R. and Hall, M.J., 1990. Screening of hydrological data: test for stationarity and relative consistency. Publ. no. 49, International Institute for Land Reclamation and Improvement (ILRI), Wageningen, The Netherlands.

Dautrebande, S., Leenaars, J.G.B., Smitz, J.S. and Vanthournout, E. (Eds.), 2000. Pilot project for the definition of environment-friendly measures to reduce the risk for flash floods in the Geul River catchment (Belgium and The Netherlands): a final report. B4-3040/97/730/JNB/C4, Technum, CSO and University of Liege.

De Bruin, H.A.R. and Stricker, J.N.M., 2000. Evaporation of grass under non-restricted soil moisture conditions. Hydrol. Sci. J., 45(3), 391–406.

De Laat, P.J.M., 2001. Workshop on hydrology (lecture note), UNESCO-IHE, Delft, The Netherlands.

De Laat, P.J.M. and Agor, M.L.C., 2003. Geen toename van piekafvoer van de Geul. H_2O, 36(9), 25–27.

De Laat, P.J.M., Calderón Hijuma, M.P. and Tu, M., 2005. Impact of human activities on the rainfall-runoff relation of the Rur catchment. In: Proceedings of 2nd International Yellow River Forum on Keeping Healthy Life of the River held at Zhengzhou, China, Yellow River Conservancy Press, Vol. III, 214–221 (also on CD-ROM).

De Laat, P.J.M. and Varoonchotikul, P., 1996. Modelling evapotranspiration of dune vegetation. In: V.P. Singh and B. Kumar (Eds.), Surface-Water Hydrology, Kluwer, 1, 19–27.

Demarée, G.R., Lachaert, P.J., Verhoeve, T. and Thoen, E., 2002. The long-term daily central Belgium temperature (CBT) series (1767–1998) and early instrumental meteorological observations in Belgium. Clim. Chan., 53(1–3), 269–293.

De Mars, H., Schuttelaar, M. and Vercoutere, B., 2000. Internationale ecologische verkenning Maas – Historisch ecologische orientatie op het stroomgebied. Rijkswaterstaat directie Limburg, The Netherlands (in Dutch).

Demuth, S., 2005. Low flows and droughts – a European perspective. IAHS Newsletter 82, March 2005. 7–8.

Demuth, S. and Stahl, K. (Eds.), 2001. ARIDE – Assessment of the regional impact of droughts in Europe. Final Report to the European Commission. Freiburg, Germany.

De Roo, A.P.J., Odijk, M., Schmuck, G., Koster, E. and Lucieer, A., 2002. Assessing the effects of land-use changes on floods in the Meuse and Oder catchment. Phys. Chem. Earth (B), 26, 593–599.

De Wit, M., Poitevin, F., Dewil, P. and De Smedt, F., 2002. A hydrological description of the Meuse basin. In: Proceedings of First International Scientific Symposium on the River Meuse. International Meuse Commission (IMC), Maastricht, The Netherlands. 19–22. http://www.riza.nl/actualiteiten/berichten/proceeding_10_12_2003.pdf.

De Wit, M., Van Deursen, W., Goudriaan, J., Boot, U., Wijbenga, A. and Van Leussen, D.W., 2003. Integrated outlook for the river Meuse (IVM): a look upstream of Borgharen. In: R.S.E.W. Leuven et al. (Eds.), Proceedings NCR-days 2002, Current themes in Dutch river research. NCR-Publ. 20-2003, Netherlands Centre for River Studies, Delft, The Netherlands, 68–69.

De Wit, M.J.M., Warmerdam, P., Torfs, P. et al., 2001. Effect of climate change on the hydrology of the river Meuse. Dutch National Research Programme on Global Air Pollution and Climate Change, Report no. 410200090, RIVM, The Netherlands.

DGRNE, 1993. Etat de l'environnement Wallon 1993 – La faune et la flore (in French). http://mrw.wallonie.be/dgrne/sibw/especes/eew/eew93/home.html

DGRNE, 1996. Etat de l'environnement Wallon 1996: paysage (in French). http://mrw.wallonie.be/dgrne/publi/etatenv/paysage/

DGRNE, 2000. Etat de l'environnement Wallon 2000: l'environnement Wallon à l'aube du XXIe siècle (approche évolutive) (in French). http://environnement.wallonie.be/eew2000/gen/framegen.htm

DGRNE, 2003. Etat de l'environnement Wallon 2003 (in French). http://environnement.wallonie.be/eew/download.asp

Douglas, E. M., Vogel R.M. and Knoll, C. N., 2000, Trends in floods and low flows in the United States, impact of spatial correlation. J. Hydrol., 240, 90–105.

Eberle, M., Buiteveld, H., Beersma, J.J., Krahe, P. and Wilke, K., 2002. Estimation of extreme floods in the river Rhine basin by combining precipitation-runoff modelling and a rainfall generator. In: M. Spreafico and R. Weingartner (Eds.), Proceedings International Conference on Flood Estimation, Berne, Switzerland. March 6–8, 2002. CHR Report II-17, Lelystad, The Netherlands, 459–467.

EEA, 1995. Europe's Environment – The Dobris assessment. Environment Report no. 1, Copenhagen. http://reports.eea.eu.int/92-826-5409-5/en

EEA, 2001. Sustainable water use in Europe. Part 3: Extreme hydrological events: floods and droughts. Environmental Issue Report no. 21, Copenhagen. http://reports.eea.eu.int/Environmental_Issues_No_21/en/enviissue21.pdf

FAO and CIFOR, 2005. Forest and floods: drowning in fiction or thriving on facts? RAP Publ. 2005/03: Forest Perspectives 2. http://www.fao.org/documents/show_cdr.asp?url_file=/docrep/008/ae929e/ae929e00.htm

Fohrer, N., Eckhardt, K., Haverkamp, S., Frede, H-G., 2001. Applying the SWAT model as decision supporting tool for land use concepts in peripheral regions in Germany. In: D.E. Stott, R.H. Mohtar and C.G. Steinhardt (Eds.), Sustaining the Global Farm. 10th International Soil Conservation Organization Meeting, May 24–29, 1999, West Lafayette, 994–999.

Folland, C.K., Karl, T.R., Christy, J.R. et al., 2001. Observed climate variability and change. In: J.T. Houghton, Y. Ding, D.J. Griggs, M. Noguer, P.J. van der Linden, X. Dai, K. Maskell, and C.A. Johnson (Eds.), Climate Change 2001: The Scientific Basis. Cambridge University Press.

Frescro, L.O., 1994. Introduction. In: L.O. Fresco *et al.* (Eds), The future of the land: mobilising and integrating knowledge for land use options. Wageningen Agricultural University, The Netherlands. John Wiley & Sons, Chichester, England, 1–8.

Frich, P., Alexander, L.V., Della-Marta, P., Gleason, B., Haylock, M., Klein Tank, A.M.G. and Peterson, T., 2002. Observed coherent changes in climatic extremes during the second half of the 20th century. Clim. Res., 19, 193–212.

García, N.O. and Mechoso, C.R., 2005. Variability in the discharge of South American rivers and in climate. Hydrol. Sci. J., 50(3), 459–478.

Gebremeskel, S., Liu, Y.B., De Smedt, F., Hoffmann, L. and Pfister, L., 2005. Assessing the hydrological effects of landuse changes using distributed modelling and GIS. Intl. J. River Basin Management, 3(1), 265–275.

Geilen, N., Pedroli, B., Van Looy, K., Krebs, L., Jochems, H., Van Rooij, S. and Van der Sluis, Th., 2001. INTERMEUSE: the Meuse reconnected. NCR-Publ. 15-2001, RIZA, Alterra, Institute for Nature Conservation and University of Metz.

Gellens D., 2000. Trend and correlation analysis of k-day extreme precipitation over Belgium. Theor. Appl. Climatol., 66(1–2), 117–129.

Gellens, D. and Roulin, E., 1998. Streamflow response of Belgian catchments to IPCC climate change scenarios. J. Hydrol., 210, 242–258.

Gellens D. and Schädler, B., 1997. Comparaison des réponses du bilan hydrique de bassins situés en Belgique et en Suisse à un changement de climat. Rev. Sci. Eau, 10(3), 395–414 (in French).

Gerstengarbe F.-W. and Werner P.C., 1999. Katalog der *Grosswetterlagen* Europas nach Paul Hess und Helmuth Brezowsky 1881–1998. Deutscher Wetterdienst, Potsdam/Offenbach am Main, Germany (in German). http://www.pik-potsdam.de/~uwerner/gwl/

Gonzalez-Hidalgo, J.C., De Luis, M., Raventos, J. and Sanchez, J.R., 2001. Spatial distribution of seasonal rainfall trends in a western Mediterranean area. Int. J. Climatol., 21, 843–860.

Gross, R., Eeles, C.W.O. and Gustard, A. (1989). Application of a lumped conceptual model to FRIEND catchments. In: International Association of Hydrological Sciences: FRIENDS in Hydrology, IAHS Publ. no. 187, Washington, 309–320.

Guzman, J.A. and Chu, M.L., 2003. SPELL-Stat statistical analysis program. Universidad Industrial de Santander, Colombia.

Hall, M.J., 2001. Statistics and stochastic processes in hydrology (lecture note), UNESCO-IHE, Delft, The Netherlands.

Helsel, D.R. and Hirsch, R.M., 1992. Statistical methods in water resources. U.S.Geological Survey: Studies in Environmental Science 49, Elsevier.

Hisdal, H., Stahl, K., Tallaksen, L.M. and Demuth, S., 2001. Have streamflow droughts in Europe become more severe or frequent? Int. J. Climatol., 21(3), 317–333.

Hisdal, H. and Tallaksen, L.M. (Eds.), 2000. Drought event definition. ARIDE Technical Report no. 6.
http://www.hydrology.uni-freiburg.de/forsch/aride/navigation/publications/pdfs/aride-techrep6.pdf

Ho, C.H., Lee, J.Y., Ahn, M.H. and Lee, H.S., 2003. A sudden change in summer rainfall characteristics in Korea during the late 1970s. Int. J. Climatol., 23, 117–128.

Hoffmann, L. and Pfister, L., 2002. FRHYMAP – Flood risk and hydrological mapping. In Hooier, A. and Van Os, A. (Eds.), Towards sustainable flood risk management in the Rhine and Meuse river basins, Final report of the IRMA-SPONGE Umbrella Program. NCR-Publ. 18-2002: IRMA-SPONGE project 03-1- 03-30.

Hooijer, A., Klijn, F., Pedroli, G., Bas, M. and Van Os, Ad.G., 2004. Towards sustainable flood risk management in the Rhine and Meuse river basins: synopsis of the findings of IRMA-SPONGE. River Res. Appl., 20(3), 343–357.

Houghton-Carr, H. (Ed.), 1999. Flood estimation handbook, 4: Restatement and application of the Flood Studies Report rainfall-runoff method. Institute of Hydrology, Wallingford, UK.

Houghton, J.T., Ding, Y., Griggs, D.J., Noguer, M., Van der Linden, P.J., Dai, X., Maskell, K. and Johnson, C.A. (Eds.), 2001. Climate Change 2001: The Scientific Basis. Contribution of Working Group I to the Third Assessment Report of the IPCC, Cambridge University Press. http://www.grida.no/climate/ipcc_tar/wg1/index.htm

Huntington, T.G., 2006. Evidence for intensification of the global water cycle: review and synthesis. J. Hydrol., 319(1–4), 83–95.

Hurrell, J.W., 1995. Decadal trends in the North Atlantic Oscillation: regional temperatures and precipitation. Science, 269(5224), 676–679.

Hurrell, J.W. and Van Loon, H., 1997. Decadal variations associated with the North Atlantic Oscillation. Clim. Chan., 36, 301–326.

ICPR, 1998. Rhine Action plan on flood defence. Koblenz.

Jaagus, J. and Ahas, R., 2000. Space-time variations of climatic seasons and their correlation with the phenological development of nature in Estonia. Clim. Res., 15, 207–219.

Jaskula-Joustra, A., 2003. River management and low flows in the river Meuse in the Netherlands. In: Proceedings of First International Scientific Symposium on the River Meuse, November 27–28, 2002, Maastricht, The Netherlands, 27–31. http://www.riza.nl/actualiteiten/berichten/proceeding_10_12_2003.pdf.

Jensen, M.E., Burman, R.D. and Allen, R.G., 1990. Evapotranspiration and irrigation water requirements. ASCE manuals and reports on engineering practice 70. ASCE, New York.

Jones, P.D., Jónsson, T. and Wheeler, D., 1997. Extension to the North Atlantic Oscillation using early instrumental pressure observations from Gibraltar and South-West Iceland. Int. J. Climatol., 17, 1433–1450.

Kaestner, W., 1997. Was macht eine Großwetterlage zur Hochwasserlage in Bayern? Deutsche Gewaesserkundliche Mitteilungen, 41(3), 107–112 (in German).

Kiersch, B., 2000. Land-use impacts on water resources: a literature review. In: Proceedings Electronic Workshop: Land-water linkages in rural watersheds, FAO Land and Water Bulletin 9, FAO, Rome.

Klein Tank, A.M.G., Wijngaard, J.B., Können, G.P. *et al.*, 2002. Daily dataset of 20th century surface air temperature and precipitation series for the European Climate Assessment. Int. J. Climatol., 22, 1441–1453.

Klein Tank, A.M.G. and Können, G.P., 2003. Trends in indices of daily temperature and precipitation extremes in Europe, 1946–99. J. Clim., 16, 3665–3680.

KNMI, 1999. De derde rapportage over de toestand van het klimaat in Nederland 1999. De Bilt, The Netherlands (in Dutch).

Können, G.P., Zaiki, M., Baede, A.P.M., Mikami, T., Jones, P.D. and Tsukuhara, T., 2003. Pre-1872 extension of the Japanese instrumental meteorological observation series back to 1819. J. Clim., 16(1), 118–131.

Kotlarski, S., Demuth, S. and Uhlenbrook, S., 2004. Der zusammenhang zwischen atmosphärischer zirkulation und niederschlag in BadenWürttemberg. Zeitschrift Hydrologie und Wasserbewirtschaftung, 48(6), 214–224 (in German).

Kruizinga, S., 1979. Objective classification of daily 500 mbar patterns. Reprinted from preprint volume: Sixth Conference on Probability and Statistics in Atmospheric Sciences, October 9–12, 1979, Banff, Alberta, Canada. American Meterological Society, Boston, Mass.

Kundzewicz, Z.W. and Robson, A.J., 2004. Change detection in hydrological records – a review of the methodology. Hydrol. Sci. J., 49(1), 7–19.

Kundzewicz, Z.W., Graczyk, D., Maurer, T., Pińskwar, I., Radziejewski, M., Svensson, C. and Szwed, M., 2005. Trend detection in river flow series: 1. Annual maximum flow. Hydrol. Sci. J., 50(5), 797–810.

La Marche, J.L. and Lettenmaier, D.P., 2000. Effects of forest roads on flood flows in the Deschutes River, Washington. Earth Surface Processes and Landforms, 26(2), 115–134.

Lambin, E.F., Geist, H.J. and Lepers, E., 2003. Dynamics of land-use and land-cover change in tropical regions. Annu. Rev. Environ. Resour., 28, 205–41.

Lammersen R., Van de Langemheen, H. and Buiteveld, H., 2000. Impact of river training and retention measures on flood conditions in the Rhine basin. In: Proceedings of European Conference on Advances in Flood Research. November 2000. PIK Report no.65, Potsdam-Institut für Klimafolgenforschung (PIK), Potsdam, Germany, 616–628.

Leander, R. and Buishand, T.A., 2004. Rainfall generator for the Meuse basin: development of a multi-site extension for the entire drainage area. KNMI-Publ. 196-III, De Bilt, The Netherlands.

Leander, R., Buishand, A., Aalders, P. and De Wit, M.J.M., 2005. Estimation of extreme floods of the river Meuse using a stochastic rainfall generator and rainfall-runoff modelling. Hydrol. Sci. J., 50(6), 1089–1103.

Leenaers, H. and Schouten, C.J., 1989. Soil erosion and floodplain soil pollution: related problems in the geographical context of a river basin. In: Proceedings of the Baltimore Symposium, May 1989, Baltimore, Maryland. IAHS Publ. no. 184, 75–83.

Lindström, G., Johansson, B., Persson, M., Gardelin, M. and Bergström, S., 1997. Development and test of the distributed HBV-96 hydrological model. J. Hydrol., 201, 272–288.

Lindström, G. and Alexandersson, H., 2004. Recent mild and wet years in relation to long observation records and future climate change in Sweden. Ambio., 33(4–5), 183–186.

Linsley, R.K., Kohler, M.A. and Paulhus, J.L.H., 1982. Hydrology for engineers (3rd ed). McGraw-Hill Series in Water Resources and Environmental Engineering, McGraw-Hill Inc.

Liu, Y.B., De Smedt, F., Hoffmann, L. and Pfister, L., 2004. Assessing land use impacts on flood processes in complex terrain by using GIS and modeling approach. Environ. Model. Assess., 9, 227–235.

Lockwood, J.G., 2001. Abrupt and sudden climatic transition and fluctuations: a review. Int. J. Climatol., 21, 1153–1179.

Lørup, J.K., Refsgaard, J.C. and Mazvimavi, D., 1998. Assessing the effect of land-use change on catchment runoff by combined use of statistical tests and hydrological modelling: case studies from Zimbabwe. J. Hydrol., 205, 147–163.

Matheussen, B., Kirschbaum, R.L., Goodman, I.A., O'Donnell, G.M. and Lettenmaier, D.P., 2000. Effects of land cover change on streamflow in the interior Columbia River basin. Hydrol. Process., 14, 867–885.

McCarthy, J.J., Canziani, O.F., Leary, N.A., Dokken, D.J. and White, K.S. (Eds.), 2001. Climate Change 2001: Impacts, Adaptation, and Vulnerability. Contribution of Working Group II to the Third Assessment Report of the IPCC, Cambridge University Press. http://www.grida.no/climate/ipcc_tar/wg2/index.htm

McCuen, R.H., 1973. The role of sensitivity analysis in hydrologic modelling. J. Hydrol., 18, 37–53.

McCuen, R.H., 2003. Modelling hydrologic change: statistical methods. Lewis Publishers, Boca Raton London, New York, Washington.

Meiners, H.G., 2002. Monitoring Garzweiler II based on the water frame directive – quantitative condition of the groundwater in the partial catchment area of the Niers. In Proceedings of First International Scientific Symposium on the River Meuse, November 27–28, 2002., Maastricht, Netherlands, 32–37. http://www.riza.nl/actualiteiten/berichten/proceeding_10_12_2003.pdf.

Mergen, P., 2002. Compared spatio-temporal distribution of the ichthyologic communities in the reservoir lakes of Nisramont (Belgium) and Esch-sur-Sûre (Grand-Duchy of Luxembourg). PhD Thesis, Unité de recherche en biologie des organismes, Namur, Belgium (in French).

Micha, J.C., and Borlee, M.C., 1989. Chapter 16. Recent historical changes on the Belgian Meuse. In: G.E. Petts, H. Moller, A.L. Roux (Eds.), Historical change of large alluvial rivers western Europe, Wiley, Chichester, 269–295.

Middelkoop, H., Van Asselt, M.B.A., Van't Klooster, S.A., Van Deursen, W.P.A., Kwadijk, J.C.J. and Buiteveld, H., 2004. Perspectives of flood management in the Rhine and Meuse rivers. River Res. Appl., 20, 327–342.

Middelkoop, H. and Van Haselen, C.O.G. (Eds.), 1999. Twice a river: Rhine and Meuse in The Netherlands. RIZA report no. 99.003, RIZA, Arnhem, The Netherlands.

Milly, P.C.D., Dunne, K.A. and Vecchia, A.V., 2005. Global patterns of trends in streamflow and water availability in a changing climate. Nature, 438, 347–350.

Milly, P.C., Wetherald, R.T., Dunne, K.A. and Delworth, T.L., 2002. Increasing risk of great floods in a changing climate. Nature, 415(6871), 514–517.

Mudelsee, M., Börngen, M., Tatzlaff, G., Grünewald, U., 2003. No upward trends in the occurrence of extreme floods in central Europe. Nature, 425(6954), 166–169.

Nachtnebel, H.P., 2003. New strategies for flood risk management after the catastrophic flood in 2002 in Europe. In: International Symposium on Integrated Disaster Risk Management: Coping with Regional Vulnerability. 3–5 July, 2003, Kyoto International Conference Hall Kyoto, Japan. http://idrm03.dpri.kyoto-u.ac.jp/Paperpdf/62nachtnebel.pdf

Nash, J.E. and Sutcliffe, J.V., 1970. River flow forecasting through conceptual models. Part I: A discussion on principles. J. Hydrol., 10, 282–290.

New, M., Todd, M., Hulme, M. and Jones, P., 2001. Review: precipitation measurements and trends in the 20th century. Int. J. Climatol., 21, 1899–1922.

Newson, M., 1995. Hydrology and the river environment (reprinted). Oxford University Press.

Niehoff, D., Fritsch, U. and Bronstert, A., 2002. Land use impacts on storm-runoff generation: scenarios of land-use change and simulation of hydrological response in a meso-scale catchment in SW-Germany. J. Hydrol., 267(1–2), 80–93.

Nonner, J.C., 2003. Introduction to hydrogeology. IHE Delft Lecture Note Series, A.A. Balkema publishers, The Netherlands.

Ott, B. and Uhlenbrook, S., 2004. Quantifying the impact of land use changes at the event and seasonal time scale using a process-oriented catchment model. Hydrol. Earth Sys. Sci., 8(1), 62–78.

Parmet, B. and Burgdorffer, M., 1995. Extreme discharges of the Meuse in the Netherlands: 1993, 1995 and 2100 – operational forecasting and long-term expectations. Phys. Chem. Earth, 20 (5–6), 485–489.

Parmet, B., Van de Langemheen, W., Chbab, E.H., Kwadijk, J.C.J., Lorenz, N.N. and Parmet, D.K., 2001. Analyse van de maatgevende afvoer van de Maas te Borgharen. RIZA Report 2002.013, RIZA, Arnhem, The Netherlands (in Dutch).

Peterson, T.C., Easterling, D.R., Karl, T.R. et al., 1998. Homogeneity adjustments of in situ atmospheric climate data: a review, Int. J. Climatol., 18(13), 1493–1517.

Pettitt, A.N., 1979. A non-parametric approach to the change-point problem. Appl. Statist., 28(2), 126–135.

Pieterse, N.M., Olde Venterink, H., Schot, P.P. and Verkroost, A.W.M., 1998. EU-LIFE project – Demonstration project for the development of integrated management plans for catchment areas of small trans-border lowland rivers: the river Dommel. 2 Streamflow, a GIS-based environmental assessment tool for lowland rivers. Utrecht University, Utrecht, The Netherlands. http://mk.geog.uu.nl/research/Dommelproject/dommel.htm

Pfister, L., Humbert, J. and Hoffmann, L., 2000. Recent trends in rainfall-runoff characteristics in the Alzette river basin, Luxembourg. Clim. Chang., 45(2), 323–337.

Pfister, L., Kwadijk, J., Musy, A., Bronstert, A. and Hoffmann, L., 2004. Climate change, Land use change and runoff prediction in the Rhine - Meuse basins. River Res. Appl., 20, 229–241.

Querner, E.P., 2000. The Effects of human intervention in the water regime. Ground Water, 38(2), 167–171.

Reed, D. (Ed.), 1999. Flood estimation handbook, 1: Overview. Institute of Hydrology. Wallingford, Oxford, UK.

Refsgaard, J.C. and Abbott, M.B., 1996. Chapter 1 The role of distributed hydrological modelling in water resources management. In: M.B. Abbott and J.C. Refsgaard (Eds.), Distributed hydrological modeling, Water Science and Technology Library, 22, Kluwer Academic.

Refsgaard, J.C. and Storm, B., 1996. Chapter 3 Construction, Calibration and Validation of hydrological models. In: M.B. Abbott and J.C. Refsgaard (Eds.), Distributed hydrological modeling, Water Science and Technology Library, 22, Kluwer Academic.

Robson, A.J., Jones, T.K., Reed, D.W. and Bayliss, A.C., 1998. A study of trend and variation in UK floods. Int. J. Climatol., 18(2), 165–182.

RWS-directie Waterhuishouding en Waterbeweging, district Zuidoost, 1984. afdeling Operationele Zaken, Stroomgebied van de Roer, Afvoeren 1953 t/m 1983. Afvoerboekje 84.2 (in Dutch).

Sahin, V. and Hall, M.J., 1996. The effects of afforestation and deforestation on water yields. J. Hydrol., 178, 293–309.

Savenije, H.H.G., 1995. Recent extreme floods in Europe and USA: challenges for the future. Phys. Chem. Earth, 20(5–6), 433–437.

Schmith, T., 2001. Global warming signature in observed winter precipitation in northwestern Europe? Clim. Res., 17, 263–274.

Schulze, R.E. and George, W.J., 1987. Simulation of effects of forest growth on water yield with a dynamic process-based user model. Geogr. Rev., 167, 575–584.

Seuna, P., 1999. Hydrological effects of forestry treatments in Finland. Invited lecture at the Int. Symp. on Flood Control, Beijing, China.

Shaw, E.M., 1991. Hydrology in practice (2nd ed). Chapman & Hall.

Slonosky, V.C., Jones, P.D. and Davies, T.D., 2000. Variability of the surface atmospheric circulation over Europe, 1774–1995. Int. J. Climatol., 20, 1875–1897.

Smakhtin, V.U., 2001. Low flow hydrology: a review. J. Hydrol., 240 (3/4), 147–186.

Sokolovskii, D.L., 1971. River runoff: theory and analysis (3rd ed.). Israel Program for Scientific Translations, Jeusalem.

Stahl, K., 2001. Hydrological drought – a study across Europe. PhD Thesis, University of Freiburg, Germany. http://www.freidok.uni-freiburg.de/volltexte/202/

Stahl, K. and Demuth, S., 1999. Linking streamflow drought to the occurrence of atmospheric circulation patterns. Hydrol. Sci. J., 44(3), 467–482.

Stam, M.H., 2002. Effects of land-use and precipitation changes on floodplain sedimentation in the nineteenth and twentieth centuries (Geul river, The Netherlands). Spec. Publs. Int. Ass. Sediment., 32, 251–267.

Stedinger, J.R., Vogel, R.M. and Georgious, E.F., 1993. Chapter 18 Frequency analysis of extreme events. In: D.R. Maidment (Ed.), Handbook of hydrology, McGraw-Hill Inc.

Steinbrich, A., Uhlenbrook, S. and Leibundgut, C., 2002. Relevance of the seasonality of precipitation and floods for flood estimation in Southwest Germany. In: Proceedings of the International Conference on Flood Estimation, March 68, 2002, Berne, Switzerland, CHRKHR Report II 17, 357–366.

Steinbrich, A., Uhlenbrook, S., Reich, T. and Kolokotronis, V., 2005. Raum-zeitliche zusammenhaenge zwischen *Grosswetterlagen* und starkniederschlaegen in Baden-Wuerttemberg. Wasserwirtschaft, 95(11), 14–49 (in German).

Stuurman, R. and Vermeulen, P., 2000. Cross border ground water flow in the Roer valley Graben. Information, Netherlands Institute of Applied Geoscience TNO – National Geological Survey 2000.

Svensson, C., Kundzewicz, Z.W. and Maurer, T., 2005. Trend detection in river flow series: 2. Flood and low-flow index series. 50(5), 811–824.

Toebes, C. and Ouryaev, V. (Eds.), 1970. Representative and experimental basins: an international guide for research and practice. UNESCO.

Tollan, A., 2002. Land-use change and floods: what do we need most, research or management? Water Sci. Technol., 45(8), 183–190.

Tomozeiu, R., Lazzeri, M. and Cacciamani, C., 2002. Precipitation fluctuations during the winter season from 1960 to 1995 over Emilia-Romagna, Italy. Theor. Appl. Climatol., 72, 221–229.

Tveito, O.E. and Ustrul, Z., 2003. A review of the use of large-scale atmospheric circulation classification in spatial climatology. Report no. 10/03 KLIMA, Norwegian Meteorological Institute, Oslo, Norway.
http://met.no/english/r_and_d_activities/publications/2003/klima-2003-10.pdf

Tu, M., Hall, M.J., De Laat, P.J.M. and De Wit, M.J.M., 2004a. Detection of long-term changes in precipitation and discharge in the Meuse basin. In: Y. Chen, K. Takara, I.D. Cluckie and F.H. De Smedt (Eds.), GIS and Remote Sensing in Hydrology, Water Resources and Environment (Proceedings of ICGRHWE held at the Three Gorges Dam, China, September 2003), IAHS Publ. no. 289, IAHS Press, Wallingford, UK, 169–177.

Tu, M., Hall, M.J., De Laat, P.J.M. and De Wit, M.J.M., 2004b. Change in the flood regime of the Meuse river under climate variability. In: B. Webb (Eds.), Hydrology: Science and Practice for the 21st Century, British Hydrological Society, London, Vol. I, 234–238.

Tu, M., Hall, M.J., De Laat, P.J.M. and De Wit, M.J.M., 2005a. Extreme floods in the Meuse river over the past century: aggravated by land-use changes? Phys. Chem. Earth, 30(4–5), 267–276.

Tu, M., De Laat, P.J.M., Hall, M.J. and De Wit, M.J.M., 2005b. Precipitation variability in the Meuse basin in relation to atmospheric circulation. Water Sci. Technol., 51(5), 5–14.

Tu, M., De Laat, P.J.M., De Wit, M.J.M. and Hall, M.J., 2005c. Change of low flows in the Meuse river during the 20th century. In: Proceedings of 2nd International Yellow River Forum on Keeping Healthy Life of the River held at Zhengzhou, China, Yellow River Conservancy Press, Vol. V, 27–37 (also on CD-ROM).

Tu, M., De Laat, P.J.M., De Wit, M.J.M. and Uhlenbrook, S., 2006. Has the risk of flooding in the Meuse increased due to climate variability and/or land use change? In: the 9th Inter-Regional Conference on Environment-Water: Concepts for Watermanagement and Multifunctional Land-Uses in Lowlands (EnviroWater2006), May 2006, Delft, The Netherlands (on CD-ROM).

Uhlenbrook, S., Mc Donnell, J. and Leibundgut, Ch., 2001. Foreword to the special issue: Runoff generation and implications for river basin modelling. Freiburger Schriften zur Hydrologie, 13, 4–13.

Uhlenbrook, S., Steinbrich, A., Tetzlaff, D. and Leibundgut, C., 2002. Regional analysis of the generation of extreme floods. In: H.A.J. Van Lanen and S. Demuth (Eds.), FRIEND 2002 – Regional Hydrology: Bridging the Gap between Research and Practice. IAHS Publ. no. 274, IAHS Press, Wallingford, UK, 243–249.

Uhlenbrook, S., 2005. Von der Abflussbildungsprozessforschung zur prozess-orientierten Modellierung – ein Review. Hydrologie und Wasserbewirtschaftung, 49(1), 13–24 (in German).

Uijlenhoet, R., De Wit, M.J.M., Warmerdam, P.M.M. and Torfs, P.J.J.F., 2001. Statistical analysis of daily discharge data of the river Meuse and its tributaries (1968–1998): assessment of drought sensitivity". Report no. 100, Wageningen University, Wageningen, The Netherlands.

Vaes, G., Willems P. and Berlamont J., 2002. 100 years of Belgian rainfall: are there trends? Water Sci. Technol., 45(2), 55–61.

Van Deursen, W., 2004. Afregelen HBV model Maasstroomgebied. Carthago Consultancy, Rotterdam (in Dutch).

Van de Griend, A. A., 1981. A weather-type hydrologic approach to runoff phenomena. PhD Thesis, Free University of Amsterdam. The Netherlands.

Van Lanen, H.A.J. and Dijksma, R., 1999. Water flow and nitrate transport to a groundwater-fed stream in the Belgian-Dutch chalk region. Hydrol. Process., 13(3), 295–307.

Van Oldenborgh, G.J., Burgers, G. and Klein Tank, A., 2000. On the EL-Niño Teleconnection to Spring precipitation in Europe. Int. J. Climatol., 20, 565–574.

Van Oldenborgh, G.J. and Van Ulden, A.A.D., 2003. On the relationship between global warming, local warming in the Netherlands and changes in circulation in the 20th century. Int. J. Climatol., 23(14), 1711–1724.

Voortman, R.L., 1998. Recent historical climate change and its effect on land-use in the eastern part of West Africa. Phys. Chem. Earth, 23(4), 385–391.

VW, 1996. Space for rivers. The Hague, The Netherlands.

VW, 2000. A different approach to water: water management policy in the 21st century. The Hague, The Netherlands.

Ward, R.C. and Robinson, M., 1990. Principles of hydrology (3rd ed). McGraw-Hill Inc., Maidenhead, UK.

Watts, G., 1996. Chapter 5 Hydrological modelling in practice. In: R.L. Wilby (Ed.), Contemporary hyrology: towards holostic environmental science, John Wiley & Sons, 151–193.

Weatherhead, E.C., Reinsel, G.C., Tiao, G.C. *et al*., 1998. Factors affecting the detection of trends: statistical considerations and applications to environmental data, J. Geophys. Res., 103(D14), 17149–17162.

Werner, P.C., Gerstengarbe, F.W., Fraedrich, K. and Oesterle, H., 2000. Recent climate change in the North Atlantic/European Sector. Int. J. Climatol., 20, 463–471.

WHM, 1998. Meuse high water action plan. Namur.

Wibig, J., 1999. Precipitation in Europe in relation to circulation patterns at the 500 hPa level. Int. J. Climatol., 19(3), 253–269.

Wijngaard, J.B., Klein Tank, A.M.G. and Konnen, G.P., 2003. Homogeneity of 20th century European daily temperature and precipitation series. Int. J. of Climatol., 23(6), 679–692.

Wind, H.G., Nierop, T.H., de Blois, C.J. and de Kok, J.L., 1999. Analysis of flood damage from the 1993 and 1995 Meuse floods. Water Resour. Res., 35(11), 3459–3465.

WL|Delft Hydraulics, 1988. Evalutie debietmeetstations waterschap "Roer en Overmaas". Report no. Q 708, Delft Hydraulics, Delft, The Netherlands (in Dutch).

WL|Delft Hydraulics, 1994. Onderzoek watersnood Maas, Deelrapport 4: Hydrologische aspecten. Delft Hydraulics, Delft, The Netherlands (in Dutch).

WMO, 1994. Guide to hydrological practices (5th ed). WMO-No.168, data acquisition and processing, analysis, forecasting and other applications.

Yarnal, B., Comrie, A.C., Frakes, B. and Brown, D.P., 2001. Review: developments and prospects in synoptic climatology. Int. J. Climatol., 21(15), 1923–1950.

Yue, S. and Pilon, P., 2003. Canadian streamflow trend detection: impacts of serial and cross-correlation. Hydrol. Sci. J., 48(1), 51–63.

Yue, S., Pilon, P. and Cavadias, G., 2002. Power of the Mann-Kendall and Spearman's rho tests for detecting monotonic trends in hydrological series. J. Hydrol., 259, 254–271.

Yue, S. and Wang, C.Y., 2002. The influence of serial correlation on the Mann-Whitney test for detecting a shift in median. Adv. Water Resour., 25(3), 325–333.

Yusuf, A., 2005. Impact of human activities on the rainfall-runoff relation of the Rur catchment using the HBV model. MSc Thesis, UNESCO-IHE, Delft, The Netherlands.

Appendix A Statistical methods

Double mass analysis

Given data sets from two stations, where X_i (i = 1, 2, ..., n) is a chronologic data set observed for a certain time length at a "reference" station and is considered to be homogeneous, where Y_i is a data set of the same variable, with the same time length, observed at another station and for which homogeneity needs to be analysed. The graphical application of the double mass analysis is done by plotting all coordinate points x_i and y_i, where x_i and y_i are the cumulative data sets created by progressively summing values of X_i and Y_i (Allen *et al.*, 1998):

$$x_i = \sum_{j=1}^{i} X_j, \quad y_i = \sum_{j=1}^{i} Y_j \quad \text{with i = 1, 2, ..., n} \tag{1-1}$$

The plot is visually analysed to determine whether successive points of x_i and y_i follow an unique straight line, which indicates the homogeneity of the data set Y_i relative to data set X_i. If there appears to be a break (or more than one break) in the plot of y_i to x_i, then there is a visual indication that the data series Y_i is not homogeneous.

The residual mass curve is often plotted against the accumulated values of each station to show more clearly the deviations of the station from the average linear relation. The residual mass is defined by (De Laat, 2001):

$$M_i = \sum_{j=1}^{i} Y_j - \left(\frac{\Sigma Y}{\Sigma X} \right) \sum_{j=1}^{i} X_j \tag{1-2}$$

where
M_i residual mass in day i of station Y
X_j daily values of station X
Y_j daily values of station Y
ΣX accumulated values of station X over the entire period considered
ΣY accumulated values of station Y over the entire period considered
i, j 1, 2, ..., n, where n is the record length

The break at coordinates x_k and y_k in the double mass plot can be used to separate two subsets (i = 1, 2, ..., k) and (k+1, k+2, ..., n). Using the meta-data or the available relevant information, the user must decide which subset requires correction. Then the following correction procedure (Allen et al., 1998) can be applied.

1) When the first subset is homogeneous

a) compute the two regression lines, with the first one through the origin

$$\hat{y}_i = b_n x_i \quad \text{with i = 1, 2 , ..., k} \tag{1-3}$$

$$\hat{y}_{nh,i} = a_{nh} + b_{nh}x_i \quad \text{with } i = k+1, k+2, \ldots, n \tag{1-4}$$

where the subscripts *h* and *nh* identify the homogenous and non-homogeneous subsets, respectively.

b) compute the differences between both regression lines for i = k+1, k+2, ..., n

$$\Delta\hat{y}_i = b_h x_i - (a_{nh} + b_{nh}x_i) \tag{1-5}$$

2) When the second subset is homogeneous

a) compute the regression line for the homogeneous subset (i = k+1, k+2, ..., n) after correcting the coordinates (x_i, y_i) using the coordinates of the break point (x_k, y_k), i.e. moving the origin of coordinates from (0, 0) to (x_k, y_k). This regression is therefore

$$\hat{y}_i = (y_k - b_h x_k) + b_h x_i \quad \text{with } i = k+1, k+2, \ldots, n \tag{1-6}$$

b) compute the regression line for the non-homogeneous subset forced to the origin

$$\hat{y}_i = b_{nh}x_i \quad \text{with } i = 1, 2, \ldots, k \tag{1-7}$$

c) compute the differences between the above regression lines

$$\Delta\hat{y}_i = [(y_k - b_h x_k) + b_h x_i] - b_{nh}x_i \tag{1-8}$$

For both 1) and 2) cases, correct the variables y_i corresponding to the non-homogeneous subset as

$$y_{c,i} = y_i + \Delta\hat{y}_i \tag{1-9}$$

where the subscript c identifies the corrected subset of the non-homogeneous subset.

Finally, compute the corrected estimates of the variables Y_i by solving the following equation for Y_i:

$$y_i = Y_i + \sum_{j=1}^{i-1} Y_i \tag{1-10}$$

Three-parameter Lognormal distribution (LN3)

A simple and efficient estimator of lower bound parameter ξ is introduced before taking logarithms (Stedinger *et al.*, 1993):

$$\xi = \frac{x_{(1)}x_{(n)} - x^2_{median}}{x_{(1)} + x_{(n)} - 2x_{median}} \tag{1-11}$$

where
$x_{(1)}$ the largest observed values
$x_{(n)}$ the smallest observed values
x_{median} the sample median

Test for Serial correlation

Given a sample of N observations, x_t, t = 1, 2, ..., N, with a mean $\bar{x}$, the serial correlation coefficients r_L are computed for lags L = 1, 2, ..., L_{max} where L_{max} should not exceed N/4 using the following simplified expression (Hall, 2001):

$$r_L = \frac{\sum_{i=1}^{N-L}(x_i - \bar{x})(x_{i+L} - \bar{x})}{\sum_{i=1}^{N}(x_i - \bar{x})^2} \tag{1-12}$$

The Anderson test (see Anderson, 1977) is used to provide the corresponding confidence limits (CLs) to r_L. The CLs for r_1 at the 95% confidence levels are estimated using a simplified formula (Hall, 2001):

$$CL = \frac{\pm 1.96}{\sqrt{N}} \tag{1-13}$$

The serial correlation is often removed from a data set using a pre-whitening approach (Burn and Hag Elnur, 2002):

$$yp_{t+1} = y_{t+1} - r_1 y_t \tag{1-14}$$

where
yp_{t+1} the pre-whitened series value for time interval t
y_t the original time series value for time interval t
r_1 the estimated first serial correlation coefficient

Spearman test for trend

List the data X_i in order, i = 1, 2, ..., n, and rank them in ascending order of magnitude, i.e. with the smallest allocated a rank r_1. For each X_i, compute $(r_i - i)$ to generate the statistic R_{sp} from the expression (Hall, 2001):

$$R_{sp} = 1 - \frac{6}{n(n^2 - 1)}\sum_{i=1}^{n}(r_i - i)^2 \tag{1-15}$$

To test the null hypothesis H_0 that there is no trend, compute the test statistic

$$t = R_{sp}\sqrt{\frac{n-2}{1 - R_{sp}^2}} \tag{1-16}$$

The t statistic is compared with the tabulated value of Student's t-distribution for a significant level of $\alpha/2$ and (n-2) degrees of freedom. If the t value computed from Eq. 1-16 is less than the tabulated value, then accept the null hypothesis H_0.

The Pettitt test

Consider a sequence of random variables $X_1, X_2, ..., X_T$, then the sequence is said to have a change point at τ if X_t for $t = 1, ..., \tau$ have a common distribution function $F_1(x)$ and X_t for $t = \tau + 1, \tau + 2, ..., T$ have a common distribution function $F_2(x)$, and $F_1(x) \neq F_2(x)$. Take the null hypothesis of "no-change" H_0: $\tau = T$, against the alternative hypothesis of "change" H_1: $1 \leq \tau < T$, a non-parametric statistic K_T for testing that the two samples $X_1, ..., X_t$ and $X_{t+1}, ..., X_T$ come from the same population is defined as (Pettitt, 1979):

$$K_T = \max_{1 \leq t < T} |U_{t,T}|$$ (1-17)

$$U_{t,T} = \sum_{i=1}^{t} \sum_{j=t+1}^{T} \text{sgn}(X_i - X_j) \text{ with sgn}(x) = 1 \text{ if } x > 0, 0 \text{ if } x = 0, -1 \text{ if } x < 0$$ (1-18)

The approximate significance probability P_{OA} associated with K_T is given by

$$P_{OA} = 2 \exp\left\{-6(K^+)^2 / (T^3 + T^2)\right\} \quad \text{for } T \to \infty$$ (1-19)

The Standard Normal Homogeneity Test (SNHT)

The test uses a statistic T_o to compare the mean of the first a years of the record with that of the last $n-a$ years. T_o will be small if the null hypothesis H_o is true, whereas large values of T_o make the H_A hypothesis more probable. A possible shift is located at the year A when T_o reaches a maximum at the year $a = A$. The test statistic T_o is defined as (Alexandersson and Moberg, 1997):

$$T_o = \max_{1 \leq a < n} T_{(a)} = \max_{1 \leq a < n} \left(a\bar{z}_1^2 + (n-a)a\bar{z}_2^2\right) \text{ with } a = 1, 2, ..., n$$ (1-20)

$$\bar{z}_1 = \frac{1}{a} \sum_{i=1}^{a} \frac{(Y_i - \bar{Y})}{s}$$ (1-21)

$$\bar{z}_2 = \frac{1}{n-a} \sum_{i=a+1}^{a} \frac{(Y_i - \bar{Y})}{s}$$ (1-22)

where
$\bar{Y}$ the mean of the sample
s the standard deviation of the sample
n the sample size

The null hypothesis H_o will be rejected, if T_0 is larger than a certain critical level tabulated in Table 1-1.

Table 1-1 Critical levels (T_{90}, T_{95} and $T_{97.5}$) of the SNHT test for single shift

n	10	20	30	40	50	60	70	80	90	100	150	250
T_{90}	5.05	6.10	6.65	7.00	7.25	7.40	7.55	7.70	7.80	7.85	8.05	8.35
T_{95}	5.70	6.95	7.65	8.10	8.45	8.65	8.80	8.95	9.05	9.15	9.35	9.70
$T_{97.5}$	6.25	7.80	8.65	9.25	9.65	9.85	10.10	10.20	10.30	10.40	10.80	11.20

Source: Alexandersson and Moberg (1997).

Appendix B Formulas

Rating curve

A rating curve gives the relation between the discharge and the gauge reading (also referred to as stage or water level reading) at a certain cross section of a river. The water level-discharge relation is often approximated with the following formula (De Laat, 2001):

$$Q = a(H - H_0)^b \qquad (2\text{-}1)$$

where
Q discharge (m^3/s)
H water level reading (m)
H_0 water level reading corresponding to zero discharge (m)
a, b constants (-)

Spatial homogeneity

In spatial homogeneity tests, data of a base station are related to data of surrounding stations. The spatial correlation of the rain-gauging stations is defined by a negative exponential function (De Laat, 2001):

$$\rho = \rho_0 e^{-\frac{r}{r_0}} \qquad (2\text{-}2)$$

where
ρ correlation at distance r (-)
ρ_0 correlation at distance 0 (-)
r distance between stations (km)
r_0 coefficient (km)
For mixed convective, orographic and advective type of rainfall, values for ρ_0 and r_0 are assumed as follows: $\rho_0 = 0.94$ and $r_0 = 300$ km.

Penman-Monteith method

The Penman-Monteith formula for potential evapotranspiration (PET) of grass with a length of 12 cm is defined as (De Laat, 2001):

$$PET_{P\text{-}M} = \frac{C}{L} \frac{sR_N + c_p\rho_a(e_a - e_d)/r_a}{s + \gamma(1 + r_c/r_a)} \qquad (2\text{-}3)$$

where
$PET_{P\text{-}M}$ potential evapotranspiration of grass (mm/day)
C constant to convert units from $kg/(m^2 \cdot s)$ to mm/day ($C = 86400$)
R_N net radiation at the earth's surface (W/m^2)
L latent heat of vaporization ($L = 2.45 \cdot 10^6$ J/kg)
s slope of the temperature-saturation vapour pressure curve (kPa/K)
c_p specific heat of air at constant pressure ($c_p = 10004.6$ J/(kg $\cdot$ K))
ρ_a density of air ($\rho_a = 1.2047$ kg/m^3 at sea level)

e_d actual vapour pressure of the air at 2 m height (kPa)
e_a saturation vapour pressure for the air temperature at 2 m height (kPa)
γ psychrometric constant ($\gamma = 0.067$ kPa/K at sea level)
r_a aerodynamic resistance (s/m)
r_c crop resistance in s/m (for grass $r_c = 70$ s/m)

The aerodynamic resistance r_a is a function of the wind speed. The following expression is used for wind velocities, U_2 (m/s), observed at a height of 2 m over grass:

$$r_a = \frac{208}{U_2} \tag{2-4}$$

Values for e_a and s may be obtained from

$$e_a = 0.6108 e^{\frac{12.27 T_a}{237.3 + T_a}} \tag{2-5}$$

$$s = \frac{4098 e_a}{(237.3 + T_a)^2} \tag{2-6}$$

where T_a is the 24 hour mean temperature of the air in °C.

The actual or dewpoint vapour pressure e_d is calculated from measurements of the relative humidity RH (%), thus

$$e_d = e_a \frac{RH}{100} \tag{2-7}$$

R_N (the net radiation) is calculated as the incoming short wave radiation at the earth's surface (or global radiation) R_S (W/m^2) minus the fraction r that is reflected and minus the net outgoing long wave radiation R_{nL} (W/m^2):

$$R_N = (1-r)R_S - R_{nL} \tag{2-8}$$

where
r reflection coefficient ($r = 0.23$ for grass) (-)
R_S $(0.20 + 0.60\,n/N) \cdot R_A$ (R_A is the short wave radiation received at the outer limits of the atmosphere; monthly averages for The Netherlands with a latitude of 52°N are given in Table 2-1.) (W/m^2)
R_{nL} $5.6745 \cdot 10^{-8} \cdot (273 + T_a)^4 \cdot (0.34 - 0.139\sqrt{e_d}) \cdot (0.1 + 0.9\,n/N)$ (W/m^2)
n/N fraction of sunshine hours per day (-)

Table 2-1 Short wave radiation R_A (W/m^2) received at the outer limits of the atmosphere

Latitude	Jan	Feb	Mar	Apr	May	Jun	Jul	Aug	Sep	Oct	Nov	Dec
52°N	91	157	252	357	440	475	457	389	292	192	111	74

Source: De Laat (2001).

Makkink formula

The Makkink formula (also known as the Radiation Method) is a simplified equation for the computation of PET based on radiation and temperature data alone. The expression is written as (De Laat, 2001):

$$PET_{Makkink} = CC_M \frac{s}{s+\gamma} \frac{R_S}{L} \qquad (2\text{-}9)$$

where
C_M 0.65 for grass in The Netherlands (-)

Gumbel (or EV1) distribution

For a given flood data series, the magnitudes of the T-year flood are estimated using the following formula (Hall, 2001):

$$Q_T = ay + c \qquad (2\text{-}10)$$

where
a scale parameter (-)
c location parameter (-)
y reduced variate for the Gumbel distribution (-)

The reduced Gumbel variate y is given by:

$$y = -\ln\left[-\ln\left(1-\frac{1}{T}\right)\right] \qquad (2\text{-}11)$$

where
T return period (years)

The location and scale parameters, c and a, can be estimated by method of moments:

$$\mu = c + 0.5572a \qquad (2\text{-}12)$$

$$\sigma^2 = \frac{\pi^2 a^2}{6} \qquad (2\text{-}13)$$

where
μ mean of the sample (-)
σ variance of the sample (-)

Model performance criteria

Nash-Sutcliffe coefficient of efficiency, *RE*, is defined by the following equation (Nash and Sutcliffe, 1970):

$$RE = 1 - \frac{\sum_{i=1}^{n}(Q_{obsi} - Q_{simi})^2}{\sum_{i=1}^{n}(Q_{obsi} - \overline{Q_{obs}})^2} \tag{2-14}$$

where

RE Nash-Sutcliffe coefficient of efficiency (-)

Q_{obsi} measured discharge at the ith time interval (m^3/s)

$\overline{Q_{obs}}$ mean of the observed discharge (m^3/s)

Q_{simi} simulated discharge at the ith time interval (m^3/s)

n number of observations (-)

The coefficient of determination, R^2, measures the degree of association between the observed and simulated discharges as estimated by the regression model and is defined by the following equation:

$$R^2 = \left[\frac{\left(n\sum_{i=1}^{n}Q_{obsi} \cdot Q_{simi}\right) - \sum_{i=1}^{n}Q_{obsi} \cdot \sum_{i=1}^{n}Q_{simi}}{\left(n\sum_{i=1}^{n}Q_{obsi}^2 - \left(\sum_{i=1}^{n}Q_{obsi}\right)^2\right)^{1/2} \cdot \left(n\sum_{i=1}^{n}Q_{simi}^2 - \left(\sum_{i=1}^{n}Q_{simi}\right)^2\right)^{1/2}} \right]^2 \tag{2-15}$$

where

R^2 coefficient of determination (-)

Other variables are as defined for Eq. 2-14.

Percentage bias, PB, measures the tendency of the simulated discharge to be larger or smaller than their observed discharge and is defined by the following equation:

$$PB = \left| \frac{\sum_{i=1}^{n}(Q_{simi} - Q_{obsi})}{\sum_{i=1}^{n}Q_{obsi}} \right| \cdot 100 \tag{2-16}$$

where

PB Percentage bias (%)

Other variables are as defined for Eq. 2-14.

About the author

Tu Min was born at Wuhan, Hubei province, China on 15 January 1969. In June 1990, she obtained her BSc degree in the Geography Department at Huazhong Normal University. Since then, she continued her study in the Pedology Department at Huazhong Agricultural University and, in June 1993, received her MSc degree with the thesis "Investigation on snail (*Oncomelania*) habitat areas using remote sensing technique (in Chinese)". From July 1993 till February 2002, she worked in the Water Resources Protection Bureau of Changjiang Water Resources Commission, where she was involved in environmental impact assessment of large-scale water projects, e.g. environmental protection planning for the resettlement areas in the Three Gorges reservoir region, environmental impact assessment of the Middle Route Project for South-to-North Water Transfer etc. In October 1998, she came to IHE at Delft, The Netherlands, as member of a "Three Lakes" group, China and, in September 1999, obtained her MEng. degree (with distinction) in Hydrology and Water Resources (groundwater hydrology). In May 2000, she came back to IHE and then carried out her MSc research work at the Dutch consultant company DHV, Amersfoort. After a half year, she received her MSc degree with the thesis "Groundwater flow modelling with conceptual model approach and automatic parameter calibration – a case study of the Mander/Getelo area". In November 2000, she went back to her organisation and worked as a senior engineer from April 2001 till she went again to UNESCO-IHE in February 2002 for starting her PhD research in surface water hydrology. She has several publications in international journals and conferences related to her research topic.

T - #0007 - 071024 - C48 - 254/178/14 [16] - CB - 9780415416948 - Gloss Lamination